Computational and Analytic Methods in Biological Sciences

Bioinformatics with Machine Learning and Mathematical Modelling

RIVER PUBLISHERS SERIES IN BIOMEDICAL ENGINEERING

Series Editor:

DINESH KANT KUMAR
RMIT University, Australia

The "River Publishers Series in Biomedical Engineering" is a series of comprehensive academic and professional books which focus on the engineering and mathematics in medicine and biology. The series presents innovative experimental science and technological development in the biomedical field as well as clinical application of new developments.

Books published in the series include research monographs, edited volumes, handbooks and textbooks. The books provide professionals, researchers, educators, and advanced students in the field with an invaluable insight into the latest research and developments.

Topics covered in the series include, but are by no means restricted to the following:

- Biomedical engineering
- Biomedical physics and applied biophysics
- Bio-informatics
- Bio-metrics
- Bio-signals
- Medical Imaging

For a list of other books in this series, visit www.riverpublishers.com

Computational and Analytic Methods in Biological Sciences
Bioinformatics with Machine Learning and Mathematical Modelling

Editors

Akshara Makrariya

VIT Bhopal University, Bhopal, India

Brajesh Kumar Jha

Pandit Deendayal Energy University- PDEU (Formerly PDPU) University in Gandhinagar, Gujarat, India

Rabia Musheer

VIT Bhopal University, Bhopal, India

Anant Kant Shukla

VIT Bhopal University, Bhopal, India

Amrita Jha

National Institute of Technology (SVNIT) Surat, India

Parvaiz Ahmad Naik

Xi'an Jiaotong University, China

River Publishers

Routledge
Taylor & Francis Group

NEW YORK AND LONDON

Published 2023 by River Publishers
River Publishers
Alsbjergvej 10, 9260 Gistrup, Denmark
www.riverpublishers.com

Distributed exclusively by Routledge
605 Third Avenue, New York, NY 10017, USA
4 Park Square, Milton Park, Abingdon, Oxon OX14 4RN

Computational and Analytic Methods in Biological Sciences / by Akshara Makrariya, Brajesh Kumar Jha, Rabia Musheer, Anant Kant Shukla, Amrita Jha, Parvaiz Ahmad Naik.

Routledge is an imprint of the Taylor & Francis Group, an informa business

ISBN 978-87-7022-695-0 (print)
ISBN 978-10-0087-987-2 (online)
ISBN 978-1-003-39323-8 (ebook master)

While every effort is made to provide dependable information, the publisher, authors, and editors cannot be held responsible for any errors or omissions.

Contents

Preface   **xiii**

List of Figures   **xv**

List of Tables   **xxi**

List of Contributors   **xxiii**

List of Abbreviations   **xxvii**

1   Modeling of Smoking Transmission Dynamics using Caputo–Fabrizio Type Fractional Derivative   **1**
Aqeel Ahmad, Muhammad Farman, Parvaiz Ahmad Naik, and Ali Akgul
   1.1   Introduction   2
   1.2   Basic Concept of Fractional Operators   4
   1.3   Model Formulation   5
   1.4   Caputo–Fabrizio Fractional Order Derivative   6
       1.4.1   Stability Analysis of Model by Using Fixed-Point Theory   8
   1.5   Numerical Results and Discussion   17
   1.6   Conclusion   20
       References   20

2   Hybrid Feature Selection Techniques Utilizing Soft Computing Methods for Classifying Microarray Cancer Data   **23**
Rabia Musheer Aziz, Amol Avinash Joshi, Kartik Kumar, and Abdu Hamid Gaani
   2.1   Introduction   24
   2.2   Proposed Framework   27
       2.2.1   Genes Extraction by ICA   27

		2.2.2	Genetic Bee Colony (GBC) Algorithm	28
	2.3	Used Classifier		28
		2.3.1	Naïve Bayes Classifier (NBC)	28
		2.3.2	Support Vector Machine (SVM) Classifier	29
	2.4	Experimental Setups		29
	2.5	Experimental Result		31
	2.6	Conclusion		35
		References		35

3 Finite Element Technique to Explicate Calcium Diffusion in Alzheimer's Disease   41

Devanshi D. Dave and Brajesh Kumar Jha

	3.1	Introduction		42
	3.2	Literature Survey		43
	3.3	Mathematical Formulations		44
		3.3.1	Calcium Buffering	44
		3.3.2	Voltage Gated Calcium Channel (VGCC)	45
		3.3.3	Endoplasmic Reticulum (ER)	46
	3.4	The Finite Element Technique		48
		3.4.1	Approximated Geometry of the Cell	49
		3.4.2	Physiological Boundary Conditions	50
		3.4.3	Meshing of the domain	51
	3.5	Results and Discussion		53
		3.5.1	For Hippocampal Neuron	53
		3.5.2	For Basal Forebrain Neuron	55
	3.6	Conclusion		57
		References		58

4 Comparative Analysis of Computational Methods used in Protein–Protein Interaction (PPI) Studies   63

Khushi Ahuja, Aditi Joshi, Navjyoti Chakraborty,
Ram Singh Purty, and Sayan Chatterjee

	4.1	Introduction		64
		4.1.1	Protein	64
		4.1.2	Protein–Protein Interaction	64
			4.1.2.1 Protein–protein interfacial characteristics	65
			4.1.2.1.1 Size and shape	65
			4.1.2.1.2 Complementarity between surfaces	65

4.1.2.1.3   Residue interface propensities .   66

4.1.2.1.4   Hydrophobicity including
Hydrogen bonding   66

4.1.2.1.5   Segmentation and secondary
structure   66

4.1.2.1.6   Conformational changes on
complex formation   67

4.1.2.2   PPI types   67

4.1.2.2.1   Homo oligomeric and Hetero
oligomeric   67

4.1.2.2.2   Obligate and non obligate
complexes   67

4.1.2.2.3   Transient and permanent
complexes   68

4.1.2.2.4   Disordered to ordered
complexes   69

4.1.2.3   PPI methods classification   69

4.2   In Silico Methods   69

4.2.1   Sequence Based Approaches   70

4.2.1.1   Ortholog based sequence approach   71

4.2.1.2   Domain pairs-based sequence approach . .   71

4.2.1.3   Statistical sequence-based approaches . .   72

4.2.1.3.1   Mirror tree method   72

4.2.1.3.2   PIPE   72

4.2.1.3.3   Co-evolutionary divergence . . .   73

4.2.1.4   Machine learning sequence-based
approaches   74

4.2.1.4.1   Auto-covariance   74

4.2.1.4.2   Pairwise similarity   74

4.2.1.4.3   Amino acid composition   75

4.2.1.4.4   Amino acid triad   75

4.2.1.4.5   UNISPPI   76

4.2.1.4.6   ETB viterbi   76

4.2.2   Structure Based Approaches   77

4.2.2.1   Template structure-based approaches . . .   77

4.2.2.1.1   PRISM   77

4.2.2.1.2   PREPPI   78

4.2.2.2   Statistical structure-based approach   78

4.2.2.2.1   PID matrix score   78

		4.2.2.2.2	Pre SPI	78
		4.2.2.2.3	Domain cohesion and coupling .	79
		4.2.2.2.4	MEGADOCK	79
		4.2.2.2.5	MetaApproach	80
	4.2.2.3	Machine learning structure-based approaches		81
		4.2.2.3.1	Random forest	81
		4.2.2.3.2	Struct2Net	81
	4.2.3	Gene Neighbourhood		82
	4.2.4	Gene Fusion		82
	4.2.5	In Silico Two-hybrid (I2h)		82
	4.2.6	Phylogenetic Tree		83
	4.2.7	Phylogenetic Profile		83
	4.2.8	Gene Expression		84
4.3	PPI Networks and Databases			85
	4.3.1	Creation of the PPI Networks		85
		4.3.1.1	Choice of databases and data selection . .	85
		4.3.1.2	Visualising PPI network	85
	4.3.2	Different Databases		86
		4.3.2.1	Interaction database	86
		4.3.2.2	Metamining databases	87
		4.3.2.3	Predictive interaction databases	88
		4.3.2.4	Pathway database	89
		4.3.2.5	Unifying database	90
4.4	Softwares Available for PPI			91
4.5	Conclusion			95
	References			95

5 Optimization of COVID-19 Risk Factors Using Fuzzy Logic Inference System

5	**Optimization of COVID-19 Risk Factors Using Fuzzy Logic Inference System**		**101**

Jyoti Gupta, Samidha Saxena, Namrata Kaushal,
and Parimeeta Chanchani

5.1	Introduction		102
5.2	Methodology		103
	5.2.1	Proposed Mamdani Fuzzy Control System	103
	5.2.2	Fuzzy Controller Design	104
	5.2.3	Parameters Identification	104
	5.2.4	Fuzzification	104
	5.2.5	Fuzzy Inference Rule Base	108

      5.2.6   Rule Evaluation By Fuzzy Inference Engine 109

      5.2.7   Defuzzification 109

  5.3   Results 110

  5.4   Discussion 113

  5.5   Conclusion and Future Work 115

      References 115

6 Dynamical Analysis of the Fractional-Order Mathematical Model of Hashimoto's Thyroiditis   119

  Neelam Singha

  6.1   Introduction 120

  6.2   Preliminaries 121

  6.3   Formulation of Fractional-Order Model of Hashimoto's Thyroiditis 122

  6.4   Stability Analysis 123

  6.5   Construction of a Numerical Solution Scheme 127

  6.6   Numerical Segment 128

  6.7   Conclusions 133

      References 133

7 Heated Laminar Vertical Jet of Pseudoplastic Fluids-Against Gravity   137

  Manisha Patel and M. G. Timol

  7.1   Introduction 137

  7.2   Basic Equations 142

  7.3   Results and Discussions 147

  7.4   Graphical Presentation 148

  7.5   Conclusion 150

      References 151

8 Analytical Solutions For Hydromagnetic Flow of Chemically Reacting Williamson Fluid Over a Vertical Cone and Wedge with Heat Source/Sink   155

  A. Subramanyam Reddy, S. Srinivas, and Anant Kant Shukla

  8.1   Introduction 156

  8.2   Formulation of the Problem 157

  8.3   Solution of the Problem 159

  8.4   Results and Discussion 160

  8.5   Conclusion 164

      References 164

**9 Aboodh Transform Homotopy Perturbation Method
for Solving Newell-Whitehead-Segel Equation   167**
Haresh P. Jani and Twinkle R. Singh
9.1   Introduction   167
9.2   Basic Definition of Aboodh Transform   168
      9.2.1   Some Properies of Aboodh Transform   169
9.3   Idea of Aboodh Transform Homotopy Perturbation
      Method   169
9.4   Some Illustrations   170
9.5   Conclusion   176
      References   177

**10 Transmission and Control of Droplet Infection from Exotic
to Native Population: A Mathematical Model   179**
Chanda Purushwani and Hema Purushwani
10.1 Introduction   180
10.2 Basic Assumptions and Formulation of the Model   181
10.3 Disease Free Equilibrium Point and Basic Reproduction
      Number   184
10.4 Stability Around the Disease Free Equilibrium Point   185
      10.4.1 Linear Stability Around the Disease Free
              Equilibrium Point E_0   186
      10.4.2 Non Linear Stability Around the Disease Free
              Equilibrium Point E_0   187
10.5 Existence of Endemic Equilibrium Point   190
10.6 Stability Around the Endemic Equilibrium Point   191
      10.6.1 Linear Stability Around the Endemic Equilibrium
              Point E_1   191
      10.6.2 Non Linear Stability Around the Endemic
              Equilibrium Point E_1   192
10.7 Sensitivity Analysis of Basic Reproduction Number and State
      Variables   196
10.8 Optimal Control Problem   196
10.9 Numerical Simulation   200
10.10 Conclusions   202
      References   206

**11 A Fractional Calculus Model to Depict the Calcium Diffusion
for Neurodegenerative Disease   209**
Hardik Joshi and Brajesh Kumar Jha

11.1 Introduction 210
11.2 Mathematical Preliminaries 211
11.3 Mathematical Modelling of Advection Diffusion Equation . 212
    11.3.1 A Fractional Calculus Model of Advection Diffusion
         Equation 214
11.4 Results and Discussion 216
11.5 Conclusion 221
    References 222

12 Approximate Analytic Solution for Tumour Growth and Human Head Heat Distribution Singular Boundary Value Model by High-Resolution Order-Preserving Fuzzy Transform
227

Navnit Jha and Kritika

12.1 Introduction 228
12.2 Preliminaries of Fuzzy Transform 230
12.3 Exponential Basis Approximated Fuzzy Components 231
12.4 Order Preserving Fuzzy Component Scheme 234
12.5 Tumor Growth and Oxygen Diffusion in a Spherical Cell
    Model 236
12.6 Heat Distribution in the Human Head 237
12.7 Numerical Simulations and Performance Evaluation 239
12.8 Conclusion and Remarks 245
    References 245

13 Analysis of One-Dimensional Groundwater Recharge by Spreading using Hybrid Differential Transform and Finite Difference Method
247

Aruna Sharma and Amit K Parikh

13.1 Introduction 248
13.2 Research Gap 249
13.3 Mathematical Formulation 250
13.4 Methodology 252
13.5 Hybrid Differential Transform and Finite Difference
    Method 253
13.6 Solution 253
13.7 Results and Discussion 255
13.8 Conclusions 257
13.9 Utilities of Research 257
    References 258

14 **Numerical Solution of Physiological Thermoregulatory Disturbances in Cold Environment** **261**
Akshara Makrariya, Rahul Makrariya, and Manisha jain
14.1 Introduction 262
14.2 Material and Methods 263
14.3 Result 266
14.4 Discussion 267
     References 267

15 **Mathematical Modelling of Transient Heat Conduction in Biological System by Finite Element Method and Coding in MATLAB** **271**
Manisha Jain and Akshara Makrariya
15.1 Introduction 271
15.2 General Procedure of Finite Element Method 272
15.3 Process of Finite Element Method 273
     15.3.1 Definition of the Problem and its Domain: One Dimensional Thermal Equation of Biological System 274
15.4 Steps Involved in Finite Element Process 275
15.5 Model-2: One Dimensional Quadratic Interpolation Model . 277
     15.5.1 Assumption of a Suitable form of Variation in T for Quadratic Element 277
15.6 Assembly of Elements 280
15.7 Matrix Form of Element Equation 280
15.8 Algorithm and Computer Program to Solve Heat Equation using FEM in Matlab: 281
15.9 Result and Discussion 287
     References 289

Index **293**

About the Editors **295**

Preface

Despite the major advances in healthcare over the past century, the successful treatment of cancer has remained a significant challenge, and cancers are the second leading cause of death worldwide behind cardiovascular disease. Early detection and survival are important issues to control cancer. The development of quantitative methods and computer technology has facilitated the formation of new models in medical and biological sciences. The application of mathematical modeling in solving many real world problems of medicine and biology has yielded fruitful results. In spite of advancements in instrumentation technology and biomedical equipment, it is not always possible to perform experiments in medicine and biology due to various reasons. Thus mathematical modeling and simulation are viewed as viable alternative in such situations. The problems of biology and medicine pose new challenges for modeling and simulation. The temperature regulation system of a human body is one of the important control systems involved in maintaining the structure and function of human body organs. The study of the temperature regulation system of human body organs is quite interesting from both application and mathematical points of view. Such models can be developed further to generate thermal information, which can be useful to biomedical scientists for the development of protocols for the diagnosis and treatment of cancer.

The conventional diagnostic techniques of cancer are not always effective as they rely on the physical and morphological appearance of the tumor. Early stage prediction and diagnosis are very difficult with conventional techniques. It is well known that cancers are involved in genome level changes. As of now, the prognosis of various types of cancer depends upon findings, related to the data generated through different experiments. Several machine-learning techniques exist in analyzing the data of expressed genes, however, the recent results related to Deep Learning algorithms are more accurate and accommodative, as it is effective in selecting and classifying informative genes. This book explores the probabilistic computational deep learning

xiii

model for cancer classification and prediction. This book is useful for an engineer, academicians, and researchers who are interested in expanding their knowledge of the applications of mathematical modeling and deep learning for a biological problem

List of Figures

Figure 1.1   Simulation of P(t) with Caputo–Fabrizio fractional order scheme.   18

Figure 1.2   Simulation of L(t) with Caputo–Fabrizio fractional order scheme.   18

Figure 1.3   Simulation of S(t) with Caputo–Fabrizio fractional order scheme.   19

Figure 1.4   Simulation of Q(t) with Caputo–Fabrizio fractional order scheme.   19

Figure 1.5   Simulation of R(t) with Caputo–Fabrizio fractional order scheme.   20

Figure 2.1   Overall flow of the proposed framework.   27

Figure 2.2   Average error rate of NB classifier with different hybrid gene selection method for the five microarray cancer datasets.   32

Figure 2.3   Average error rate of SVM classifier with different hybrid gene selection method for the five cancer microarray datasets.   33

Figure 2.4-2.8 Classification accuracy of NB and SVM classifiers with different number of genes with proposed approach (ICA+GBC) for above define five data sets.   34

Figure 3.1   Calcium diffusion in presence of buffers   47

Figure 3.2   Approximated structure of a hippocampal neuron .   50

Figure 3.3   Approximated structure of a basal forebrain neuron   50

Figure 3.4   Single and multiple boundary conditions in hippocampal neuron   51

Figure 3.5   Single and multiple boundary conditions in basal forebrain neuron.   51

Figure 3.6   Initial and refined meshing of the hippocampal neuron   52

Figure 3.7   Initial and refined meshing of the basal forebrain neuron   52

Figure 3.8   Triangle quality for initial and refined mesh in hippocampal neuron   53

Figure 3.9   Triangle quality for initial and refined mesh in basal forebrain neuron   53

Figure 3.10   Calcium concentration distribution for case 1 and case 2 in normal and Alzheimeric cells having single flux   54

Figure 3.11   Calcium concentration distribution for case 1 and case 2 in normal and Alzheimeric cells having multiple fluxes   55

Figure 3.12   Calcium concentration distribution for case 1 and case 2 in normal and Alzheimeric cells having single flux   56

Figure 3.13   Calcium concentration distribution for case 1 and case 2 in normal and Alzheimeric cells having multiple fluxes.   57

Figure 4.1   Types of protein–protein interactions   68

Figure 4.2   Binary and Co-complex method for protein–protein interaction (adapted from Rivas and Fontanillo, 2010). Here, X represents the interaction that does not take place (i.e., they are false positive results) .   70

Figure 5.1   Fuzzy Control System   104

Figure 5.2   Membership function for Body Temperature in three linguistic terms   105

Figure 5.3   Membership function for Cough in two linguistic terms   106

Figure 5.4   Membership function for Cold in two linguistic terms   106

Figure 5.5   Membership function for Breathing Problem in three linguistic terms   107

Figure 5.6   Membership function for Loss Sense of Smell/Taste in three linguistic terms   108

Figure 5.7   Fuzzy Inference System   110

Figure 5.8   Membership function for Risk% in three linguistic terms   111

Figure 5.9   3D form Rule Surface   112

Figure 6.1   Plot of the variable $x_1(t, \alpha)$ = TSH(t, α) (Thyroid Stimulating Harmone). 132

Figure 6.2   Plot of the variable $x_2(t, \alpha)$ = FT 4(t, α) (Free Thyroxin). 132

Figure 6.3   Plot of the variable $x_3(t, \alpha)$ = Ab(t, α) (Anti-Thyroid Antibodies). 133

Figure 6.4   Plot of the variable $x_4(t, \alpha)$ = T (t, α) (Functional Size of the Thyroid Gland). 133

Figure 7.1   Garden Fountain. 138

Figure 7.2   Volcanic Irruption. 138

Figure 7.3   Blood flow through heart. 139

Figure 7.4   Co-ordinate system for two-dimensional vertical jet 143

Figure 7.5   Velocity Profile for different values of Pr. 148

Figure 7.6   Velocity profile for different values of Gr. 148

Figure 7.7   Velocity profile for different values of n for Pr=50 & Gr=10. 149

Figure 7.8   Temperature Profile for different values of Gr. . . . 149

Figure 7.9   Temperature profile for different values of Pr. . . . 150

Figure 8.1   The coordinate system of the model. 157

Figure 8.2   Velocity profiles (a) impact of γ (b) impact of M (c) impact of Q (d) impact of K. 161

Figure 8.3   Temperature profiles (a) impact of Q (b) impact of M (c) impact of m_1 (d) impact of Gr_x. 162

Figure 8.4   Concentration profiles (a) impact of K (b) impact of m_2 (c) impact of Gc_x. 163

Figure 9.1   Approximate solution for E.g.1 172

Figure 9.2   Approximate solution for E.g.1 172

Figure 9.3   Approximate solution for E.g.1 173

Figure 9.4   Exact solution for E.g. 2 174

Figure 9.5   Approximate solution for E.g. 3 176

Figure 9.6   Exact solution for E.g. 3 176

Figure 10.1   Disease propagation sketch of designed model . . . 183

Figure 10.2   Graph between Time versus Infected population and Pathogens with one control out of three controls. 200

Figure 10.3   Graph between Time versus All Population in disease free and endemic states. 202

Figure 10.4   Graph between Time versus Infected population and Pathogens for various of (a).   203

Figure 10.5   Graph between Time versus Infected population and Pathogens for various values of (β_2).   203

Figure 10.6   Graph between Time versus Infected population and Pathogens for various values of (η).   204

Figure 10.7   Graph between Time versus Infected population and Pathogens for Various values of (δ).   204

Figure 10.8   Graph between Pathogens versus Infected Population   205

Figure 10.9   Graph between Basic Reproduction Number versus $(a, \ \beta_2, \delta, \eta)$.   205

Figure 11.1   Graph of calcium profile against time for various fractional order derivative in presence of advection flux.   217

Figure 11.2   Graph of calcium profile against time for various fractional order derivative in absence of advection flux.   217

Figure 11.3   Graph of calcium profile against time for various amounts of advection flux at $\alpha = 1$.   218

Figure 11.4   Graph of calcium profile against distance for various fractional order derivative in presence of advection flux.   219

Figure 11.5   Graph of calcium profile against distance for various fractional order derivative in absence of advection flux.   219

Figure 11.6   Graph of calcium profile against distance for various amounts of advection flux at $\alpha = 1$. . . .   220

Figure 11.7   Graph of calcium profile against time for various value of diffusion coefficients at $\alpha = 1$.   220

Figure 11.8   Graph of calcium profile against distance for various value of diffusion coefficients at $\alpha = 1$.   221

Figure 12.1   Effect of parameter μ on the solution.   238

Figure 12.2   Impact of parameter η on the solution.   238

Figure 12.3   Solution behavior for changing d.   240

Figure 12.4   Solution behavior for changing σ.   240

Figure 12.5   Effect of ν on the solution errors   242

Figure 12.6   Graph of frequency parameter versus errors.   244

Figure 12.7   Impact of frequency parameter on errors.   244

Figure 13.1   Ground water recharge   248
Figure 13.2   Moisture Content Vs. Depth for fixed time T   256
Figure 13.3   Moisture Content Vs. Depth and Time   256
Figure 15.1   One dimensional simplex elements for
interpolation   276
Figure 15.2   (a) 1-D Linear Element (b) Discretization of
domain using One Dimensional Linear Element . .   277
Figure 15.3   (a) 1-D Quadratic Element (b) Discretization of the
domain using One Dimensional Quadratic
Element   278
Figure 15.4   Graph between temperatures T and time t for $T_a =$
$15\,°C$, E=0.0 gm/cm^2-min and $\eta=\theta=0.01$.   288
Figure 15.5   Graph between temperatures T and time t
for T_a=15 $°C$, E=0.0 gm/cm^2-min and $\eta=\theta=0.01$. .   288

List of Tables

Table 2.1   Summary of five benchmark microarray gene expression datasets. 29

Table 2.2   Classification accuracy of NB and SVM classifiers with best selected gene sets by using different feature selection technique for colon cancer data 30

Table 2.3   Classification accuracy of NB and SVM classifiers with best selected gene sets by using different feature selection technique for colon cancer data for Acute leukemia data 30

Table 2.4   Classification accuracy of NB and SVM classifiers with best selected gene sets by using different feature selection technique for colon cancer data for Prostate tumor data. 30

Table 2.5   Classification accuracy of NB and SVM classifiers with best selected gene sets by using different feature selection technique for colon cancer data for High-grade Glioma data. 31

Table 2.6   Classification accuracy of NB and SVM classifiers with best selected gene sets by using different feature selection technique for colon cancer data for Lung cancer II data. 31

Table 3.1   Value of physiological parameters 54

Table 4.2   Links for different databases 90

Table 4.3   PPI Prediction tools 91

Table 4.4   Protein binding site prediction tools 92

Table 4.5   Protein–Protein Prediction tools for model organisms   93

Table 4.6   Structure Based PPI interaction tool 93

Table 4.7   PPI Tools for network creation 94

Table 4.8   Other PPI tools 94

Table 5.1   Parameters and membership values with linguistic ranges 104

Table 5.2   Fuzzy logic-based rules   109

Table 5.3   Risk% of the infected patient   111

Table 6.1   Estimated values of parameters   128

Table 6.2   Standard values and range of the variables x_i, $1{\leq}i{\leq}4$.   129

Table 8.1   Comparison of present numerical values of $f''(0)$ and $\theta'(0)$ with Chamkha and Vajravelu and Nayfeh for $n = 1, \gamma = 0$, $\mathrm{Pr} = 0.3$, $Q = -5, Gr_x = -0.5, M = 1, Gc_x = K = Sc = m_2 = 0.$ 160

Table 8.2   Effects of γ, M, K, Q, Gr_x, Gc_x on Nu and Sh. . . .   163

Table 10.1   Sensitivity indices of R_0 with respect to parameter x_i   196

Table 10.2   Sensitivity indices of state variables with respect to parameter   197

Table 10.3   Basic Reproductive Number (R_0) and Equilibrium points for various values of (β_2)   201

Table 10.4   Basic Reproductive Number (R_0) and Equilibrium points for various values of (a)   201

Table 10.5   Basic Reproductive Number (R_0) and Equilibrium points for various values of (η)   201

Table 10.6   Basic Reproductive Number (R_0) and Equilibrium points for various values of (δ)   201

Table 12.1   Integrated absolute errors and convergence order in example 12.1.   242

Table 12.2   Integrated absolute errors and convergence order at $\tau = 6$ in example 12.2.   242

Table 12.3   Maximum absolute and root-mean-squared errors in example 12.3.   243

Table 12.4   Errors and convergence order in example 12.4 for changing ν.   243

Table 12.5   Integrated absolute errors and convergence order in example 12.5.   243

Table 13.1   Moisture content θ Vs Depth Z at fixed time T   255

List of Contributors

Ahmad Naik, Parvaiz, *School of Mathematics and Statistics, Xi'an Jiaotong University, Xi'an, Shaanxi, 710049, P. R. China*

Ahmad, Aqeel, *Department of Mathematics and Statistics, University of Lahore, Lahore-54590, Pakistan*

Ahuja, Khushi, *University School of Biotechnology, Guru Gobind Singh Indraprastha University, New Delhi, India*

Akgul, Ali, *Art and Science Faculty, Department of Mathematics, Siirt University, 56100 Siirt Turkey*

Aziz, Rabia Musheer, *Department of SASL (Mathematics)*

Chakraborty, Navjyoti, *University School of Biotechnology, Guru Gobind Singh Indraprastha University, New Delhi, India*

Chanchani, Parimeeta, *Department of Engineering Science & Humanities, Indore Institute of Science & Technology, Indore, MP, India*

Chatterjee, Sayan, *University School of Biotechnology, Guru Gobind Singh Indraprastha University, New Delhi, India; E-mail: sayan@ipu.ac.in*

Dave, Devanshi D., *Pandit Deendayal Energy University; E-mail: ddave1822@gmail.com*

Farman, Muhammad, *Department of Mathematics and Statistics, University of Lahore, Lahore-54590, Pakistan*

Gaani, Abdu Hamid, *Saudi Electronic University*

Gupta, Jyoti, *Department of Engineering Science \& Humanities, Indore Institute of Science & Technology, Indore, MP, India; E-mail: jyoti.gupta@indoreinstitute.com*

Jain, Manisha, *Vellore Institute of Technology, Bhopal University E-mail: mujain31@gmail.com; Manisha.jain@vitbhopal.ac.in*

Jani, Haresh P., *Applied Mathematics and Humanities Department, Sardar Vallabhbhai National Institute of Technology, Surat-395007 (Gujarat), India; E-mail: hareshjani67@gmail.com*

Jha, Brajesh Kumar, *Department of Mathematics, School of Technology, Pandit Deendayal Energy University, Gandhinagar-382007; E-mail: brajeshjha2881@gmail.com*

Jha, Navnit, *Faculty of Mathematics and Computer Science South Asian University, Delhi, India; E-mail: navnitjha@sau.ac.in*

Joshi, Amol Avinash, *Department of SASL (Mathematics)*

Joshi, Aditi, *University School of Biotechnology, Guru Gobind Singh Indraprastha University, New Delhi, India*

Joshi, Hardik, *Department of Mathematics, LJ Institute of Engineering and Technology, LJ University, Ahmedabad-382210; E-mail: hardik.joshi8185@gmail.com*

Kaushal, Namrata, *Department of Engineering Science \& Humanities, Indore Institute of Science & Technology, Indore, MP, India*

Kritika, *Faculty of Mathematics and Computer Science South Asian University, Delhi, India*

Kumar, Kartik, *Department of SASL (Mathematics); Department of Computer Seience, VIT Bhopal University, Bhopal- Indore Highway, Kothrikalan, Sehore, -466116 (M.P.) India*

Makrariya, Rahul, *Sagar Institute of Research and Technology, Bhopal-MP, India*

Makrariya, Akshara, *School of Advanced Science- Mathematics, VIT University Bhopal; E-mail: aksharahul@gmail.com*

Parikh, Amit K., *Principal, Mehsana Urban Institute of Sciences, Ganpat University,Kherva-384012, Mehsana, Gujarat, India*

Patel, Manisha, *Department of Mathematics, Sarvajanik College of Engineering \& Technology, Surat-395001, Gujarat, India; E-mail: manishapramitpatel@gmail.com*

Purty, Ram Singh, *University School of Biotechnology, Guru Gobind Singh Indraprastha University, New Delhi, India*

Purushwani, Chanda, *Department of Mathematics, School of Science, ITM University, Gwalior 474010, MP, India;*
E-mail: chandapurushwani1987@outlook.com

Purushwani, Hema, *Department of Mathematics, School of Science, ITM University, Gwalior 474010, MP, India; E-mail: hemapurushwani1985@outlook.com*

Reddy, A. Subramanyam, *Department of Mathematics, School of Advanced Sciences, VIT, Vellore- 632014, Tamil Nadu, India;*
E-mail: anala.subramanyamreddy@gmail.com

Saxena, Samidha, *Department of Engineering Science \& Humanities, Indore Institute of Science \& Technology, Indore, MP, India*

Sharma, Aruna, *Research Scholar, Department of Mathematics, Mehsana Urban Institute of Sciences,Ganpat University, Kherva-384012, Mehsana, Gujarat, India; E-mail: harmaaruna961@gmail.com*

Shukla, Anant Kant, *Department of Mathematics, School of Advanced Sciences and Languages, VIT Bhopal University, Sehore – 466114, India*

Singh, Twinkle R., *Applied Mathematics and Humanities Department, Sardar Vallabhbhai National Institute of Technology, Surat-395007 (Gujarat), India; E-mail: winklesingh.svnit@gmail.com*

Singha, Neelam, *Department of Mathematics, School of Technology, Pandit Deendayal \\Energy University, Gandhinagar, Gujarat-382426, India*

Srinivas, S., *Department of Mathematics, School of Sciences and Languages, VIT-AP University, Amaravati - 522 237, India*

Timol, M. G., *Retired Professor, Department of Mathematics, Veer Narmad South Gujarat University, Magdalla Road, Surat-395007, Gujarat, India; E-mail: mgtimol@gmail.com*

List of Abbreviations

α_i	Permanent recovery rate from infected population to recovered population
β_i	Rate of infection transmits to susceptible human being's through pathogens $(i = 1, 2)$
Δ_1	Native human being's recruitment rate
Δ_2	Exotic human being's recruitment rate
δ	Pathogen's natural death rate
η	Pathogen's produce rate by infected human being
μ	Human being's natural death rate
ϑ_i	Temporary recovery rate from recovered population to susceptible population
σ_i	Human being's disease caused death rate
a	Efficiency of mask used by population
$[B]_\infty$ Calmodulin	Buffer concentration
$[B]_\infty$ Calbindin	Buffer concentration
$[Ca^{2+}]_\infty$	Background calcium concentration
C_0	Total calcium concentration
C_1	Volume ratio
D_{leak}	Diffusivity constant
D_{chan}	Conductance of the channel
k^+ Calmodulin	Buffer association rate
k^+ Calbindin	Buffer association rate
$P_R{}^{max}$	Maximum pump rate
P_{oout}	Rate of calcium efflux from cytosol
Z_{Ca}	Valency of calcium ion
ADM	Adomian decomposition method
D	Diffusion coefficient
DTM	Differential transform method
EADM	Elzaki Adomian decomposition method
F	Faraday's constant
GA	Genetic algorithm
GBC	Genetic bee colony

HAM	Homotopy analysis method
HAM	Homotopy analysis method
HATM	Homotopy analysis transform Method
HDTFD	Hybrid differential transform finite difference
HPM	Homotopy perturbation method
HPM	Homotopy perturbation method
ICA	Independent component analysis
K	Michaelis-Menten constant
MDTM	Modified differential transform method
NB	Naïve Bayes
ODE	Ordinary differential equation
PDE	Partial differential equations
PSO	particle swarm optimization
R	Ideal gas constant
RDTM	Reduced differential transform method
SVM	Support vector machine
T	Temperature
TDDTM	Two-Dimensional differential transform method
V	Volume of cytosol
VIM	Variational iteration method

1

Modeling of Smoking Transmission Dynamics using Caputo–Fabrizio Type Fractional Derivative

Aqeel Ahmad[1], Muhammad Farman[1], Parvaiz Ahmad Naik[2,*], and Ali Akgul[3]

[1]Department of Mathematics and Statistics, University of Lahore, Lahore-54590, Pakistan
[2]School of Mathematics and Statistics, Xi'an Jiaotong University, Xi'an, Shaanxi, 710049, P. R. China
[3]Art and Science Faculty, Department of Mathematics, Siirt University, 56100 Siirt Turkey
*Corresponding Author

Abstract

In this chapter, the true purpose of the current work is to develop and analyze the Caputo–Fabrizio scheme of fractional derivative model for the Smoking epidemic, which contains an antiretroviral treatment compartment. Stability analysis has been made using self-mapping and Banach space and derives a unique solution for the fractional-order model, which is a new approach for such type of biological models. A few numerical simulations are done by using the given method of fractional order to explain and support the theoretical results.

Keywords: Epidemic smoking model, Sumudu Transform, Caputo–Fabrizio, Uniqueness.

1.1 Introduction

Mathematics was firstly used in biology in the twelfth century when Fibonacci used his popular Fibonacci series to explain a growing population. Daniel Bernoulli used mathematics to describe the effect of small forms. The term biological mathematics was primarily used by Johannes Reinke in 1901. It is aimed at the mathematical image and modeling of biological processes. It is also used to recognize phenomena in the living organism. Biomathematics has made major progress during the last few decades, and this progress will continue in upcoming decades. Math has played a large role in natural science, but now it also will be more useful in biology. We should teach basic concepts of biomathematics at the early stages. The basic steps in mathematical biology are few. The initial step is to explain the biological process and raise questions about the basics. The second step is to build up a mathematical model that represents the underlying biological process. The 3rd step is to apply the method, and math concepts have to predict the model. The last step is to check whether this prediction gives answers to the raised questions. After this, anyone can further explore the biological question by using mathematical models [1].

Now in the modern world, tobacco smoke is the most inhaled substance. Tobacco is made by mixing its agricultural form with many substances. The smoke is inhaled through the lungs. The most dangerous epidemic in the world is the smoking epidemic. Due to smoking, 50% of its users died. Every year about 60 million people die due to smoking. During the last few decades, it has seen a huge boost in deaths. The death rate will rise thrice annually in 2030, almost seventy percent is in developing countries. According to WHO, 10 million people will die due to smoking. As compared to other diseases, the death ratio is higher than all. The person who uses tobacco dies 14 years earlier than those who do not smoke [2]. Tobacco smoking is the major reason of cancer and other diseases. About 70% of people died due to tobacco-related diseases in developing countries [3]. Three million people die due to smoking yearly. Now a day's smoking is the most dangerous habit. The heart attack ratio is 70 percent more than a nonsmoker. There are 900 million men smokers and 200 women smokers in the world. After every 6 seconds, there is a death due to smoking. Smoking is a major cause of lung and heart attacks in the world.

Due to smoking, the chances of other diseases like heart attack, stroke, especially lung cancer, Throat, mouth, esophagus, and pancreas are increased tobacco causes many tissues related diseases. Tobacco smoke is a mixture of several toxic gases. It includes 98 of which are linked with an increased risk of cardiovascular disease, 69 of which are carcinogenic. Daily a smoker takes 1 or 2 milligrams of nicotine per cigarette. Thus, if a person smokes 5 cigarettes, it takes at least 5–10 milligrams of nicotine. The effect of smoking is not limited to a person but also has adverse effects on other people. It causes 22% of death annually.

The chemical composition of tobacco varies according to the environment. These leaves are mixed with many chemicals. Tobacco smoke contains many different chemicals such as benzopyrene, NNK, aldehydes, carbon monoxide, hydrogen cyanide, phenol, nicotine, and harmalaalkoids. The radioactive element polonium 210 is also occurring in tobacco. There are almost 4 thousand noxious substances in smoke which is the main reason for cancer [4, 5]. The chemical composition of smoke depends on puff frequency and other materials. Nicotine is the main issue for disturbing the nervous system, rise in heartbeat, raising blood pressure, and shrinking the small blood vessels, which are the main basis of wrinkles. The amount of oxygen decreased in the lungs due to carbon monoxide (CO). The natural lung cleaner that is minuscule hairs is destroyed by hydrogen cyanide. Lead, nickel, arsenic, and cadmium are also present in the smoke. Some pesticides like DDT are also found in smoke. The major reason for skin and lung cancer is toxic chemicals that do present in smoke.10 million deaths will occur in the 20th century and 1 billion in the 21st century due to smoking [6]. Cigarette smoking badly affects human fertility [7]. The chemical is a major reason which affects testicular function and spermatogenesis. The smoke can affect females during egg production and pregnancy. Smoking affects almost every organ of the human body. Due to tobacco, the chance of heart attack is increased, it is also a major cause of heart diseases. Smoking during teenage can increase the risk of COPD.

NRT, varenicline, and bupropion these treatments can be used for the treatment of smoking. NRT is a treatment in which a low amount of nicotine without other poisonous chemicals is given to the patient. NRT takes 8 to 12 weeks. Varenicline can be used in two ways. It lessens the craving for nicotine. Its treatment also takes about 12 weeks. Originally it was used for depression, but it is also useful in quitting smoking. 1 to 2 tablets are used daily. Its treatment takes around 7 to 9 weeks.

The generalization of classical calculus is called fractional calculus, which is concerned with the integration and differentiation of fractional order. In the 19th century by using fractional calculus, mathematicians introduced fractional differential equations, fractional dynamics, and fractional geometry. Fractional calculus is used in almost every field of science. It is used to model physical as well as engineering processes .in many cases, standard mathematical models of integer order do not work properly. Due to this reason, fractional calculus made a major contribution to mechanics, chemistry, biology, and image processing. By using fractional calculus, several physical problems are solved. By using integer-order derivatives the system shows many problems such as history and nonlocal effect.

Primarily, all the studies depended on Caputo fractional-order and Reimann-Liouville fractional derivatives. Now a day it has been highlighted that these derivatives have the issue, and the issue is they have a singular kernel. That is the reason so many new definitions were presented in the studies [8–16]. These new definitions were very impactful because they have nonsingular kernels which are according to their needs. Caputo fractional derivatives [17], the Caputo–Fabrizio derivative [11], and Atangna–Baleanue [18] fractional derivative differ from each other only because Caputo is defined by a power law, Fabrizio is defined by using exponential decay law, and Atangna–Beleanue is defined by Mittage–Leffler law. Tateishi et al. describe the role of fractional time operator derivative in a study of anomalous diffusion [19]. With the help of analytical techniques Bulut et al. deliberate the role of differential equations of arbitrary order [20]. The key concepts of fractional differential equation and their application are explained by Kilbas et al. [21]. Antangna and Alkahtani examined the Keller–Segel model about a fractional derivative having a nonsingular kernel [22]. Fractional logistic maps are newly introduced by Huang et al. [23]. Zaman studied the qualitative response of the dynamics of giving up smoking [24]. The giving up smoking model linked with Caputo fractional derivative is a probe by Ertuk et al. [25].

The conversion of daily life problems into mathematical models is called mathematical modeling. It is the branch of mathematical logic that facilitates to structure of any daily life problem into mathematical equations, which is also called mathematical modeling and then finding the solution.

1.2 Basic Concept of Fractional Operators

Definition 1.2.1: For a function $g(t) \in W_2^1(0,1)$, $b > a$ and $\sigma \in [0,1]$, the definition of Atangana–Baleanu derivative in the Caputo sense is given by

$$^{ABC}_{0}D_t^\sigma g(t) = \frac{AB(\sigma)}{1-\sigma} \int_0^t \frac{d}{d\tau} g(\tau) M_\sigma \left[-\frac{\sigma}{1-\sigma}(t-\tau)^\sigma \right] d\tau, \quad n-1 < \sigma < n$$

$$(1.1)$$

where

$$AB(\sigma) = 1 - \sigma + \frac{\sigma}{\tau(\sigma)}.$$

By using Sumudu transform (ST) for (11), we obtain

$$ST[^{ABC}_0 D^\sigma_t g(t)](s) = \frac{q(\sigma)}{1 - \sigma}\left\{\sigma\tau(\sigma + 1)M_\sigma\left(-\frac{1}{1 - \sigma}V^\sigma\right)\right\}$$
$$\times [ST(g(t)) - g(0)]. \tag{1.2}$$

Definition 1.2.2: The Laplace transform of the Caputo fractional derivative of a function $g(t)$ of order $\sigma > 0$ is defined as

$$L[^C_0 D^\sigma_t g(t)] = s^\sigma g(s) - \sum_{\sigma=0}^{n-1} g^{(\sigma)}(0)s^{\sigma-v-1}. \tag{1.3}$$

Definition 1.2.3: The Laplace transform of the function $t^{\sigma_1-1}E_{\sigma,\sigma_1}(\pm\mu t^\sigma)$ is defined as

$$L[t^{\sigma_1-1}E_{\sigma,\sigma_1}(\pm\mu t^\sigma)] = \frac{s^{\sigma-\sigma_1}}{s^\sigma \mp \mu}, \tag{1.4}$$

Where E_{σ,σ_1} is the two-parameter Mittag–Leffler function with $\sigma, \sigma_1 > 0$. Further, the Mittag–Leffler function satisfies the following equation [17].

$$E_{\sigma,\sigma_1}(f) = fE_{\sigma,\sigma+\sigma_1}(f) + \frac{1}{\Gamma(\sigma_1)}. \tag{1.5}$$

Definition 1.2.4: Suppose that $g(t)$ is continuous on an open interval (a, b), then the fractal-fractional integral of $g(t)$ of order σ having Mittag–Lefflertype kernel and given by

$$^{FFM}_\square J^{\sigma,\sigma_1}_{0,t}(g(t)) = \frac{\sigma\sigma_1}{AB(\sigma)\tau(\sigma_1)}\int_0^t s^{\sigma_1-1}g(s)(t-s)^\sigma ds$$
$$+ \frac{\sigma_1(1-\sigma)t^{\sigma_1-1}g(t)}{AB(\sigma)}. \tag{1.6}$$

1.3 Model Formulation

We will study the giving up the smoking model probe by Ertuk et al. (1.25). T(t) is the overall population at time t. We separate the population into 5 groups, potential smokers P(t), occasional smokers L(t), heavy smokers S(t), temporary quitters Q(t), and smokers who quit permanently R(t) specified by T(t)=P(t)+L(t)+S(t)+Q(t)+R(t). The model is developed as follows

$$\frac{dP}{dt} = a\left(1 - P\right) - bPS$$

$$\frac{dL}{dt} = -aL + bPL - cLS$$

$$\frac{dS}{dt} = -\left(a + d\right) + cLS + fQ \qquad (1.7)$$

$$\frac{dQ}{dt} = -\left(a + f\right)Q + d\left(1 - e\right)S$$

$$\frac{dR}{dt} = -aR + edS$$

The comes with an initial condition

$$P\left(0\right) = \delta_1, L\left(0\right) = \delta_2, S\left(0\right) = \delta_3, Q\left(0\right) = \delta_4, R\left(0\right) = \delta_5$$

In scheme (1.7), the connection between potential smokers and occasional smokers is represented by b, the rate of natural death among occasional and temporary smokers, the connection between quitters and smokers, is shown by f, and the rate of giving up smoking is shown by d, the fraction of temporary giving up smoker is represented by (1–e) (at the rate of d), e shows the remaining fraction of smokers who give up smoking forever (at a rate d).

Now,

1.4 Caputo–Fabrizio Fractional Order Derivative

$$^{CF}_{0}D^{\rho}_{t}P = a\left(1 - P\right) - bPS$$

$$^{CF}_{0}D^{\rho}_{t}L = -aL + bPL - cLS$$

$$^{CF}_{0}D^{\rho}_{t}S = -\left(a + d\right)S + cLS + fQ \qquad (1.8)$$

$$^{CF}_{0}D^{\rho}_{t}Q = -\left(a + f\right)Q + d\left(1 - e\right)S$$

$$^{CF}_{0}D^{\rho}_{t}R = -aR + fdS$$

Applying Sumudu Transform on (1.8)

$$ST\left(^{CF}_{0}D^{\rho}_{t}P\right) = ST\left(a(1 - P)\right) - bPS)$$

$$ST\left({}^{CF}_0 D_t^\rho L\right) = ST(-aL + bPL - cLS)$$

$$ST\left({}^{CF}_0 D_t^\rho S\right) = ST(-(a+d)\,S + cLS + fQ)$$

$$ST\left({}^{CF}_0 D_t^\rho Q\right) = ST(-(a+f)\,Q + d\,(1-e)\,S)$$

$$ST\left({}^{CF}_0 D_t^\rho R\right) = ST(-aR + edS)$$

$$\frac{M\,(\rho)\,ST\,[P(t) - P(0)]}{1 - \rho + \rho\mu} = ST\,[a(1 - P) - bPS]$$

$$\frac{M\,(\rho)\,ST\,[L\,(t) - L\,(0)]}{1 - \rho + \rho\mu} = ST\,[-aL + bPL - cLS]$$

$$\frac{M\,(\rho)\,ST\,[S\,(t) - S\,(0)]}{1 - \rho + \rho\mu} = ST\,[-(a+d)\,S + cLS + fQ] \qquad (1.9)$$

$$\frac{M\,(\rho)\,ST\,[Q\,(t) - Q\,(0)]}{1 - \rho + \rho\mu} = ST\,[-(a+f)\,Q + d\,(1-e)\,S]$$

$$\frac{M\,(\rho)\,ST\,[R(t) - R(0)]}{1 - \rho + \rho\mu} = ST\,[-aR + edS]$$

Rearranging equation (1.9)

$$ST\,(P\,(t)) = P\,(0) + \frac{1 - \rho + \rho\mu}{M\,(\rho)} ST\,[a(1 - P) - bPS]$$

$$ST\,(L\,(t)) = L\,(0) + \frac{1 - \rho + \rho\mu}{M\,(\rho)} ST\,[-aL + bPL - cLS]$$

$$ST\,(S\,(t)) = S\,(0) + \frac{1 - \rho + \rho\mu}{M\,(\rho)} ST\,[-(a+d)\,S + cLS + fQ] \quad (1.10)$$

$$ST\,(Q\,(t)) = Q\,(0) + \frac{1 - \rho + \rho\mu}{M\,(\rho)} ST\,[-(a+f)\,Q + d\,(1-e)\,S]$$

$$ST\,(R\,(t)) = R\,(0) + \frac{1 - \rho + \rho\mu}{M\,(\rho)} ST\,[-aR + edS]$$

Now taking Inverse Sumudu Transform on (1.10)

$$P\,(t) = P\,(0) + ST^{-1}\left\{\frac{1 - \rho + \rho\mu}{M\,(\rho)} ST\,[a(1 - P) - bPS]\right\}$$

$$L\,(t) = L\,(0) + ST^{-1}\left\{\frac{1 - \rho + \rho\mu}{M\,(\rho)} ST\,[-aL + bPL - cLS]\right\}$$

$$S\,(t) = S\,(0) + ST^{-1}\left\{\frac{1 - \rho + \rho\mu}{M\,(\rho)} ST\,[-(a+d)\,S + cLS + fQ]\right\}$$

$$Q(t) = Q(0) + ST^{-1}\left\{ \frac{1-\rho+\rho\mu}{M(\rho)} ST\left[-(a+f)Q + d(1-e)S \right] \right\}$$

$$R(t) = R(0) + ST^{-1}\left\{ \frac{1-\rho+\rho\mu}{M(\rho)} ST\left[-aR + edS \right] \right\}$$

We must obtain the following recursive Formula

$$P_{n+1}(t) = P_n(0) + ST^{-1}\left\{ \frac{1-\rho+\rho\mu}{M(\rho)} ST\left[a(1-P_n) - bP_nS_n \right] \right\}$$

$$L_{n+1}(t) = L_n(0) + ST^{-1}\left\{ \frac{1-\rho+\rho\mu}{M(\rho)} ST\left[-aL_n + bP_nL_n - cL_nS_n \right] \right\}$$

$$S_{n+!}(t) = S_n(0)$$

$$+ ST^{-!}\left\{ \frac{1-\rho+\rho\mu}{M(\rho)} ST\left[-(a+d)S_n + cL_nS_n + fQ_n \right] \right\}$$

$$\tag{1.11}$$

$$Q_{n+1}(t) = Q_n(0)$$

$$+ ST^{-1}\left\{ \frac{1-\rho+\rho\mu}{M(\rho)} ST\left[-(a+f)Q_n + d(1-e)S_n \right] \right\}$$

$$R_{n+1}(t) = R_n(0) + ST^{-1}\left\{ \frac{1-\rho+\rho\mu}{M(\rho)} ST\left[-aR_n + edS_n \right] \right\}$$

Solution of (1.11) is

$$P(t) = \lim_{n\to\infty} P_n(t)$$

$$L(t) = \lim_{n\to\infty} L_n(t)$$

$$S(t) = \lim_{n\to\infty} s_n(t)$$

$$Q(t) = \lim_{n\to\infty} Q_n(t)$$

$$R(t) = \lim_{n\to\infty} R_n(t)$$

1.4.1 Stability Analysis of Model by Using Fixed-Point Theory

Theorem 1.4.1.1: Suppose that,$(X_1, \|.\|)$ as a Banach space and P a self-map of X_1 satisfying $\|P_x - P_y\| \le C\|X - P_x\| + c\|x - y\|$ for all $x, y \in X_1$, and $0 \le C, 0 \le c < 1$. Assume that P is Picard P-Stable. We consider the system (21), we have

$$P_{n+1}(t) = P_n(0) + ST^{-1}\left\{\frac{1-\rho+\rho\mu}{M(\rho)}ST\left[a(1-P_n)-bP_nS_n\right]\right\}$$

$$L_{n+1}(t) = L_n(0) + ST^{-1}\left\{\frac{1-\rho+\rho\mu}{M(\rho)}ST\left[-aL_n+bP_nL_n-cL_nS_n\right]\right\}$$

$$S_{n+!}(t) = S_n(0)$$
$$+ ST^{-!}\left\{\frac{1-\rho+\rho\mu}{M(\rho)}ST\left[-(a+d)S_n+cL_nS_n+fQ_n\right]\right\}$$

$$Q_{n+1}(t) = Q_n(0)$$
$$+ ST^{-1}\left\{\frac{1-\rho+\rho\mu}{M(\rho)}ST\left[-(a+f)Q_n+d(1-e)S_n\right]\right\}$$

$$R_{n+1}(t) = R_n(0) + ST^{-1}\left\{\frac{1-\rho+\rho\mu}{M(\rho)}ST\left[-aR_n+edS_n\right]\right\}$$

Where $\frac{1-\rho+\rho\mu}{M(\rho)}$ is the fractional Lagrange multiplier.

Theorem 1.4.1.2 Let us define a self-map as

$$P\left(P_n(t)\right) = P_{n+1}(t) = P_n(0)$$
$$+ ST^{-1}\left\{\frac{1-\rho+\rho\mu}{M(\rho)}ST\left[a(1-P_n)-bP_nS_n\right]\right\}$$

$$P\left(L_n(t)\right) = L_{n+1}(t) = L_n(0)$$
$$+ ST^{-1}\left\{\frac{1-\rho+\rho\mu}{M(\rho)}ST\left[-aL_n+bP_nL_n-cL_nS_n\right]\right\}$$

$$P\left(S_n(t)\right) = S_{n+!}(t) = S_n(0)$$
$$+ ST^{-!}\left\{\frac{1-\rho+\rho\mu}{M(\rho)}ST\left[-(a+d)S_n+cL_nS_n+fQ_n\right]\right\}$$

$$P\left(Q_n(t)\right) = Q_{n+1}(t) = Q_n(0)$$
$$+ ST^{-1}\left\{\frac{1-\rho+\rho\mu}{M(\rho)}ST\left[-(a+f)Q_n+d(1-e)S_n\right]\right\}$$

$$P\left(R_n(t)\right) = R_{n+1}(t)$$
$$= R_n(0) + ST^{-1}\left\{\frac{1-\rho+\rho\mu}{M(\rho)}ST\left[-aR_n+edS_n\right]\right\}$$

Is P-stable in $L^1(a, b)$ *if*

$$
\begin{cases}
\{1 - af(r) - b(M + L)g(r)\} < 1 \\
\{1 - af(r) - b(M + L_1)g_1(r) - c(M_1 + L)h(r)\} < 1 \\
\{1 - (a + d)f_1(r) + fg_2(r) + c(M_1 + L)h(r)\} & < 1 \\
\{1 - (a + f)f_2(r) + d(1 - e)g_3(r)\} < 1 \\
\{1 - af(r) + edg_4(r)\} < 1
\end{cases}
$$

$$(1.12)$$

Proof: First of all, we will show that P has a fixed point. For this, we evaluate the following for all $(m \times n) \in N \times N$

$$
P((P_n t)) - P(P_m(t)) = (P_n(t)) - (P_m(t))
$$
$$
+ ST^{-1}\left\{\frac{1 - \rho + \rho\mu}{M(\rho)} ST[a(1 - P_n) - bP_n S_n]\right\}
$$
$$
- ST^{-1}\left\{\frac{1 - \rho + \rho\mu}{M(\rho)} ST[a(1 - P_m) - bP_m S_m]\right\}
$$
$$
P(L_n(t)) - P(L_m(t)) = L_n(t) - L_m(t)
$$
$$
+ ST^{-1}\left\{\frac{1 - \rho + \rho\mu}{M(\rho)} ST[-aL_n + bP_n L_n - cL_n S_n]\right\}
$$
$$
- ST^{-1}\left\{\frac{1 - \rho + \rho\mu}{M(\rho)} ST[-aL_m + bP_m L_m - cL_m S_m]\right\}
$$
$$
P(S_n(t)) - P(S_m(t)) = S_n(t) - S_m(t)
$$
$$
+ ST^{-!}\left\{\frac{1 - \rho + \rho\mu}{M(\rho)} ST[-(a + d)S_n + cL_n S_n + fQ_n]\right\}
$$
$$
- ST^{-!}\left\{\frac{1 - \rho + \rho\mu}{M(\rho)} ST[-(a + d)S_m + cL_m S_m + fQ_m]\right\}
$$

$$(1.13)$$

$$
P(Q_n(t)) - P(Q_m(t)) = Q_n(t) - Q_m(t)
$$
$$
+ ST^{-1}\left\{\frac{1 - \rho + \rho\mu}{M(\rho)} ST[-(a + f)Q_n + d(1 - e)S_n]\right\}
$$
$$
- ST^{-1}\left\{\frac{1 - \rho + \rho\mu}{M(\rho)} ST[-(a + f)Q_m + d(1 - e)S_m]\right\}
$$
$$
P(R_n(t)) - P(R_m(t)) = R_n(t) - R_m(t)
$$

$$+ ST^{-1} \left\{ \frac{1 - \rho + \rho\mu}{M(\rho)} ST [-aR_n + edS_n] \right\}$$

$$- ST^{-1} \left\{ \frac{1 - \rho + \rho\mu}{M(\rho)} ST [-aR_m + edS_m] \right\}$$

We take the first equation of the system of (1.13) and by using the norm on both hand sides, we have

$$\left\| P((P_n t)) - P(P_m(t)) \right\| = \left\| (P_n(t)) - (P_m(t)) \right.$$

$$+ ST^{-1} \left\{ \frac{1 - \rho + \rho\mu}{M(\rho)} ST [a(1 - P_n) - bP_n S_n] \right\}$$

$$- ST^{-1} \left\{ \frac{1 - \rho + \rho\mu}{M(\rho)} ST [a(1 - P_m) - bP_m S_m] \right\} \right\|$$

$$\left\| P(L_n(t)) - P(L_m(t)) \right\| = \left\| L_n(t) - L_m(t) \right.$$

$$+ ST^{-1} \left\{ \frac{1 - \rho + \rho\mu}{M(\rho)} ST [-aL_n + bP_n L_n - cL_n S_n] \right\}$$

$$- ST^{-1} \left\{ \frac{1 - \rho + \rho\mu}{M(\rho)} ST [-aL_m + bP_m L_m - cL_m S_m] \right\} \right\|$$

$$\left\| P(S_n(t)) - P(S_m(t)) \right\| = \left\| S_n(t) - S_m(t) \right.$$

$$+ ST^{-!} \left\{ \frac{1 - \rho + \rho\mu}{M(\rho)} ST [-(a + d) S_n + cL_n S_n + fQ_n] \right\}$$

$$- ST^{-!} \left\{ \frac{1 - \rho + \rho\mu}{M(\rho)} ST [-(a + d) S_m + cL_m S_m + fQ_m] \right\} \right\|$$

$$\left\| P(Q_n(t)) - P(Q_m(t)) \right\| = \left\| Q_n(t) - Q_m(t) \right.$$

$$+ ST^{-1} \left\{ \frac{1 - \rho + \rho\mu}{M(\rho)} ST [-(a + f) Q_n + d(1 - e) S_n] \right\}$$

$$- ST^{-1} \left\{ \frac{1 - \rho + \rho\mu}{M(\rho)} ST [-(a + f) Q_m + d(1 - e) S_m] \right\} \right\|$$

$$\left\| P(R_n(t)) - P(R_m(t)) \right\| = \left\| R_n(t) - R_m(t) \right.$$

$$+ ST^{-1} \left\{ \frac{1 - \rho + \rho\mu}{M(\rho)} ST [-aR_n + edS_n] \right\}$$

$$-ST^{-1}\left\{\frac{1-\rho+\rho\mu}{M(\rho)}\,ST\,[-aR_m+edS_m]\right\}\bigg\| \qquad (1.14)$$

We write the equation (1.14) in the triangular inequality form and then we have

$$\|P((P_n t))-P(P_m(t))\| \le \|(P_n(t))-(P_m(t))\|$$
$$+\left\|ST^{-1}\left\{\frac{1-\rho+\rho\mu}{M(\rho)}ST\,[a(1-P_n)-bP_nS_n]\right\}\right.$$
$$-ST^{-1}\left\{\frac{1-\rho+\rho\mu}{M(\rho)}ST\,[a(1-P_m)-bP_mS_m]\right\}\bigg\| \qquad (1.15)$$

Upon further simplification, (1.15) yields that

$$\|P((P_n t))-P(P_m(t))\| \le \|(P_n(t))-(P_m(t))\|$$
$$+ST^{-1}\left|ST\frac{1-\rho+\rho\mu}{M(\rho)}\right|\|-a(P_n-P_m)\|$$
$$+\|-bP_n(S_n-S_m)\|+\|-bS_m(P_n-P_m)\| \qquad (1.16)$$

Since both sides of the solution play the same rule, so we shall assume

$$\|(P_n(t))-(P_m(t))\| \cong \|L_n(t)-L_m(t)\|$$
$$\|(P_n(t))-(P_m(t))\| \cong \|S_n(t)-S_m(t)\|$$
$$\|(P_n(t))-(P_m(t))\| \cong \|Q_n(t)-Q_m(t)\|$$
$$\|(P_n(t))-(P_m(t))\| \cong \|R_n(t)-R_m(t)\|$$

Replacing in equation (1.16) we obtain

$$\left\|P((P_n t))-P(P_m(t))\right\| \le \|(P_n(t))-(P_m(t))\|$$
$$+ST^{-1}\left|ST\frac{1-\rho+\rho\mu}{M(\rho)}\right|\|-a(P_n-P_m)\|$$
$$+\|-bP_n(P_n-P_m)\|+\|-bS_m(P_n-P_m)\| \qquad (1.17)$$

Since P_n and S_m are bounded as they are convergent sequences, therefore we can find different positive constants M and L for all such that

$$\|P_n\| \le M, \|S_m\| \le L \ (m \times n) \in N \times N \qquad (1.18)$$

Now considering equation (1.17) with (1.18), we get

$$\|P((P_n t))-P(P_m(t))\| \le \left[1+ST^{-1}\left\{ST\frac{1-\rho+\rho\mu}{M(\rho)}\right\}\|-a\|\right.$$

$$+ST^{-1}\left\{ST\frac{1-\rho+\rho\mu}{M\left(\rho\right)}\right\}\left\|-bS_m\right\|$$

$$+ST^{-1}\left\{ST\frac{1-\rho+\rho\mu}{M\left(\rho\right)}\right\}\left\|-bP_n\right\|\right]\left\|P_n-P_m\right\| \tag{1.19}$$

$$\left\|P\left(\left(P_n t\right)\right)-P\left(P_m\left(t\right)\right)\right\|$$
$$\leq\left[1+ST^{-1}\left\{ST\frac{1-\rho+\rho\mu}{M\left(\rho\right)}\right\}\left\|-a\right\|\right.$$
$$\left.+ST^{-1}\left\{ST\frac{1-\rho+\rho\mu}{M\left(\rho\right)}\right\}\left\|-b(P_{n+}S_m)\right\|\right]\left\|P_n-P_m\right\|$$

$$\left\|P\left(\left(P_n t\right)\right)-P\left(P_m\left(t\right)\right)\right\|$$
$$\leq\left\{1-af\left(r\right)-b\left(M+L\right)g\left(r\right)\right\}\left\|P_n-P_m\right\| \tag{1.20}$$

Where f, are functions from $ST^{-1}\left\{ST\frac{1-\rho+\rho\mu}{M(\rho)}\right\}$
In the same way, we get

$$\left\|P\left(\left(L_n t\right)\right)-P\left(L_m\left(t\right)\right)\right\|$$
$$\leq\left\{1-af\left(r\right)-b\left(M+L_1\right)g_1\left(r\right)-c\left(M_1+L\right)h(r)\right\}\left\|L_n-L_m\right\| \tag{1.21}$$

$$\left\|P\left(\left(S_n t\right)\right)-P\left(S_m\left(t\right)\right)\right\|$$
$$\leq\left\{1-\left(a+d\right)f_1\left(r\right)+fg_2\left(r\right)+c\left(M_1+L\right)h(r)\right\}\left\|S_n-S_m\right\| \tag{1.22}$$
$$\left\|P\left(\left(Q_n t\right)\right)-P\left(Q_m\left(t\right)\right)\right\|$$
$$\leq\left\{1-\left(a+f\right)f_2\left(r\right)+d(1-e)g_3\left(r\right)\right\}\left\|Q_n-Q_m\right\| \tag{1.23}$$
$$\left\|P\left(\left(R_n t\right)\right)-P\left(R_m\left(t\right)\right)\right\|$$
$$\leq\left\{1-af\left(r\right)+edg_4\left(r\right)\right\}\left\|R_n-R_m\right\| \tag{1.24}$$

Where

$$\left\{1-af\left(r\right)-b\left(M+L\right)g\left(r\right)\right\}<1$$
$$\left\{1-af\left(r\right)-b\left(M+L_1\right)g_1\left(r\right)-c\left(M_1+L\right)h(r)\right\}<1$$
$$\left\{1-\left(a+d\right)f_1\left(r\right)+fg_2\left(r\right)+c\left(M_1+L\right)h(r)\right\}<1$$
$$\left\{1-\left(a+f\right)f_2\left(r\right)+d(1-e)g_3\left(r\right)\right\}<1$$
$$\left\{1-af\left(r\right)+edg_4\left(r\right)\right\}<1.$$

Hence proved P has a fixed point. We also prove that it fulfills all conditions of 4.1.2.

$$c = (0, 0, 0, 0, 0)$$

$$= c \begin{cases} \{1 - af(r) - b(M + L)g(r)\} \\ \{1 - af(r) - b(M + L_1)g_1(r) - c(M_1 + L)h(r)\} \\ \{1 - (a + d)f_1(r) + fg_2(r) + c(M_1 + L)h(r)\} \\ \{1 - (a + f)f_2(r) + d(1 - e)g_3(r)\} \\ \{1 - af(r) + edg_4(r)\} \end{cases}$$

Hence proved that P is Picard P-stable.

Theorem 1.4.1.3: Prove that system (1.8) has a special solution that is unique.

Proof: Suppose that Hilbert space $H = L((a, b) \times (0, T))$ which is given as

$$y : (a, b) \times (0, T) \to \mathbb{R}, \iint uy \, du \, dy < \infty.$$

Suppose that $P(P, L, S, Q, R) = \begin{cases} a(1 - P) - bPS \\ -aL + bPL - cLS \\ -(a + d)S + cLS + fQ \\ -(a + f)Q + d(1 - e)S \\ -aR + edS \end{cases}$

The inner product is the major purpose of this part

$P((X_{11} - X_{12}, X_{21} - X_{22}, X_{31} - X_{32}, X_{41} - X_{42}, X_{51} - X_{52})$,
$(W_1, W_2, W_3, W_4, W_5))$
$((X_{11} - X_{12}), (X_{21} - X_{22}), (X_{31} - X_{32}), (X_{41} - X_{42})$,
$(X_{51} - X_{52}),)$ is the special system of the solution, however
$P((X_{11} - X_{12}, X_{21} - X_{22}, X_{31} - X_{32}, X_{41} - X_{42}, X_{51} - X_{52})$,
$(W_1, W_2, W_3, W_4, W_5))$

$$= \begin{cases} \{-a(X_{11} - X_{12}) - b(X_{11} - X_{12})(X_{31} - X_{32}), W_1\} \\ \{-a(X_{21} - X_{22}) + b(X_{11} - X_{12})(X_{21} - X_{22}) - c(X_{21} - X_{22})(X_{31} - X_{32}), W_2\} \\ \{-(a + d)(X_{31} - X_{32}) + c(X_{21} - X_{22})(X_{31} - X_{32}) + f(X_{41} - X_{42}), W_3\} \\ \{-(a + f)(X_{41} - X_{42}) + d(1 - e)(X_{31} - X_{32}), W_4\} \\ \{-a(X_{51} - X_{52}) + ed(X_{31} - X_{32}), W_5\} \end{cases}$$

We shall evaluate the first equation in the system without loss of generality

$$-a(X_{11} - X_{12}) - b(X_{11} - X_{12})(X_{31} - X_{32})$$

$$\cong (-a\,(X_{11} - X_{12})\,, W_1) + (-b\,(X_{11} - X_{12})\,(X_{31} - X_{32})\,, W_1) \quad (1.25)$$

Since both solutions play the same role, we can assume that

$$(X_{11} - X_{12}) \cong (X_{21} - X_{22}) \cong (X_{31} - X_{32}) \cong (X_{41} - X_{42})$$
$$\cong (X_{51} - X_{52})$$

Then

$$(-a\,(X_{11} - X_{12}) - b(X_{11} - X_{12})^2, W_1)$$
$$\leq a\,\|X_{11} - X_{12}\|\,\|W_1\| + b\,\left\|(X_{11} - X_{12})^2\right\|\,\|W_1\|$$
$$= (a + b\overline{\omega_1})\,\|X_{11} - X_{12}\|\,\|W_1\| \quad (1.26)$$

Repeating the same practice, from the 2nd, 3rd, 4th and 5th equations of the system, we can obtain as following

$$-a\,(X_{21} - X_{22}) + b\,(X_{11} - X_{12})\,(X_{21} - X_{22}) - c\,(X_{21} - X_{22}) \times$$
$$(X_{31} - X_{32}) \leq (a + b\overline{\omega}_2 + c\overline{\omega}_2)\,\|(X_{21} - X_{22})\|\,\|w_2\|$$
$$(1.27)$$

$$-(a + d)\,(X_{31} - X_{32}) + c\,(X_{21} - X_{22})\,(X_{31} - X_{32}) + f\,(X_{41} - X_{42})$$
$$\leq (a + d + c\overline{\omega}_3 + f)\,\|(X_{31} - X_{32})\|\,\|w_3\|$$
$$(1.28)$$

$$-(a + f)\,(X_{41} - X_{42}) + d\,(1 - e)\,(X_{31} - X_{32}) \leq (a + f + d\,(1 - e))$$
$$\|(X_{41} - X_{42})\|\,\|w_4\|$$
$$(1.29)$$

$$-a\,(X_{51} - X_{52}) + ed\,(X_{31} - X_{32}) \leq (a + ed)\,\|(X_{51} - X_{52})\|\,\|w_5\|$$
$$(1.30)$$

Putting equations (1.26–1.30) in (1.25) we get

$$P\,((X_{11} - X_{12}, X_{21} - X_{22}, X_{31} - X_{32}, X_{41} - X_{42}, X_{51} - X_{52})\,,$$
$$(W_1, W_2, W_3, W_4, W_5)) \quad (1.31)$$

$$\leq \begin{cases} (a + b\overline{\omega_1})\,\|X_{11} - X_{12}\|\,\|W_1\| \\ (a + b\overline{\omega}_2 + c\overline{\omega}_2)\,\|(X_{21} - X_{22})\|\,\|w_2\| \\ (a + d + c\overline{\omega}_3 + f)\,\|(X_{31} - X_{32})\|\,\|w_3\| \\ (a + f + d\,(1 - e))\,\|(X_{41} - X_{42})\|\,\|w_4\| \\ (a + ed)\,\|(X_{51} - X_{52})\|\,\|w_5\| \end{cases} \quad (1.32)$$

But, for sufficiently large values of m_i, with $i = 1, 2, 3, 4, 5$ both the solutions converge to the exact solution, using the topological concept, there exist five very small positive parameters $l_{m_1}, l_{m_2}, l_{m_3}, l_{m_4}$ and l_{m_5} such that

$$\|(P - X_{11})\|, \|(P - X_{12})\| < \frac{l_{m_1}}{5\,(a + b\overline{\omega_1})\,\|X_{11} - X_{12}\|\,\|W_1\|},$$

$$\|(L - X_{21})\|, \|(L - X_{22})\| < \frac{l_{m_2}}{5\,(a + b\overline{\omega}_2 + c\overline{\omega}_2)\,\|(X_{21} - X_{22})\|\,\|w_2\|},$$

$$\|(S - X_{31})\|, \|(S - X_{32})\| < \frac{l_{m_3}}{5\,(a + d + c\overline{\omega}_3 + f)\,\|(X_{31} - X_{32})\|\,\|w_3\|},$$

$$\|(Q - X_{41})\|, \|(Q - X_{42})\| < \frac{l_{m_4}}{5\,(a + f + d\,(1 - e))\,\|(X_{41} - X_{42})\|\,\|w_4\|},$$

$$\|(R - X_{51})\|, \|(R - X_{52})\| < \frac{l_{m_5}}{5\,(a + ed)\,\|(X_{51} - X_{52})\|\,\|w_5\|}.$$

Thus, by taking the exact solution on the right-hand side of the equation (1.32) and applying the triangular inequality by taking $M = \max\,(m_1, m_2, m_3, m_4, m_5)$, $l = \max\,(l_{m_1}, l_{m_2}, l_{m_3}, l_{m_4}, l_{m_5})$. We obtain

$$\begin{cases} (a + b\overline{\omega_1})\,\|X_{11} - X_{12}\|\,\|W_1\| \\ (a + b\overline{\omega}_2 + c\overline{\omega}_2)\,\|(X_{21} - X_{22})\|\,\|w_2\| \\ (a + d + c\overline{\omega}_3 + f)\,\|(X_{31} - X_{32})\|\,\|w_3\| \\ (a + f + d\,(1 - e))\,\|(X_{41} - X_{42})\|\,\|w_4\| \\ (a + ed)\,\|(X_{51} - X_{52})\|\,\|w_5\| \end{cases} < \begin{cases} l \\ l \\ l \\ l \\ l \end{cases}$$

As l is a small positive parameter, therefore, based on the topological idea, we have

$$\begin{cases} (a + b\overline{\omega_1})\,\|X_{11} - X_{12}\|\,\|W_1\| \\ (a + b\overline{\omega}_2 + c\overline{\omega}_2)\,\|(X_{21} - X_{22})\|\,\|w_2\| \\ (a + d + c\overline{\omega}_3 + f)\,\|(X_{31} - X_{32})\|\,\|w_3\| \\ (a + f + d\,(1 - e))\,\|(X_{41} - X_{42})\|\,\|w_4\| \\ (a + ed)\,\|(X_{51} - X_{52})\|\,\|w_5\| \end{cases} < \begin{cases} 0 \\ 0 \\ 0 \\ 0 \\ 0 \end{cases}$$

But it is obvious that

$$(a + b\overline{\omega_1}) \neq 0$$
$$(a + b\overline{\omega}_2 + c\overline{\omega}_2) \neq 0$$
$$(a + d + c\overline{\omega}_3 + f) \neq 0$$

$$(a + f + d(1 - e)) \neq 0$$
$$(a + ed) \neq 0$$

Therefore, we have

$$\|(X_{11} - X_{12})\| = 0,$$
$$\|(X_{21} - X_{22})\| = 0,$$
$$\|(X_{31} - X_{32})\| = 0,$$
$$\|(X_{41} - X_{42})\| = 0,$$
$$\|(X_{51} - X_{52})\| = 0.$$

This yields that

$$X_{11} = X_{12}, X_{21} = X_{22}, X_{31} = X_{32}, X_{41} = X_{42}, X_{51} = X_{52}$$

This shows that the special solution is unique.

1.5 Numerical Results and Discussion

A mathematical study of the non-linear epidemic model of smoking has been presented. For checking of parameters effects on the smoking dynamical model, some numerical simulations according to the value of the parameters are accomplished to confirm the effect of the fractional derivative on the different compartments. We got mathematical consequences of the model for different fractional values with the help of Caputo–Fabrizio. If we note the impacts of variables on the dynamics of the model of fractional-order, the end-time value of the given parameter can be observed in various numerical ways. We can observe that the results of fractional values are more accurate as compared to classical derivatives. Desired results can be achieved to analyze the epidemic that occurs due to smoking. The graphs of the approximate solutions are given in Figures 1.1–1.5 against different fractional-order α. P(t) and L(t) starts decreasing by decreasing the fractional values, while S(t), Q(t), and R(t) start increasing by decreasing fractional values, which can be easily observed in Figures 1.1–1.5. When the fractional values decrease, then the behavior approaches steady-state in all figures, which shows that the solution will be more effective by decreasing the fractional values.

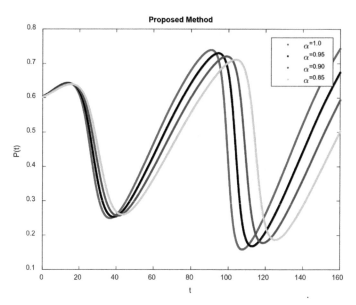

Figure 1.1   Simulation of P(t) with Caputo–Fabrizio fractional order scheme.

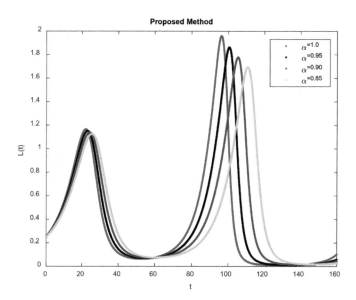

Figure 1.2   Simulation of L(t) with Caputo–Fabrizio fractional order scheme.

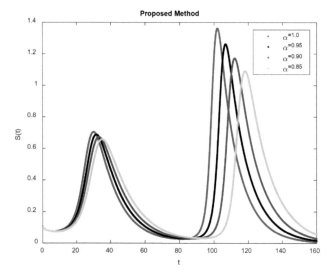

Figure 1.3 Simulation of S(t) with Caputo–Fabrizio fractional order scheme.

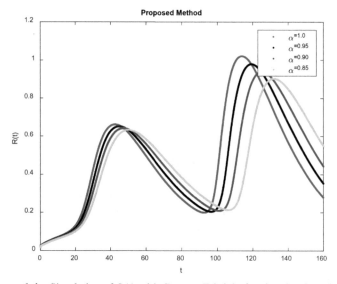

Figure 1.4 Simulation of Q(t) with Caputo–Fabrizio fractional order scheme.

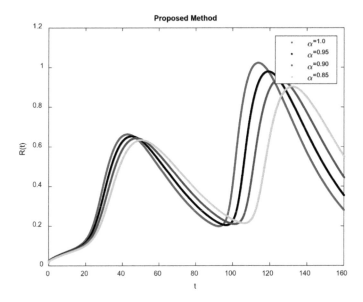

Figure 1.5   Simulation of R(t) with Caputo–Fabrizio fractional order scheme.

1.6 Conclusion

In the sense of Caputo–Fabrizio, the new numerical scheme of the fractional differential equation model has been investigated in this article for Smoking with an antiretroviral treatment section. With the help of the fixed-point theory of uniqueness and stability of the smoking, the model has been examined. The arbitrary derivative of fractional order has been taken in the Caputo–Fabrizio sense with no singular kernel. Results are obtained by using Sumudu transform on the fractional-order model. We obtained very effective results for the proposed model. Also discussed are some theoretical results that proved the efficiency of the proposed techniques. Numerical simulations are carried out to check the actual behavior of the dynamic. These results will be helpful to understand further analysis and controlling this outbreak.

References

[1] Chou, C. S., Friedman A. (2016) Introduction. In: Introduction to Mathematical Biology. Springer Undergraduate Texts in Mathematics and Technology. Springer, Cham. https://doi.org/10.1007/978-3-319-29 638-8_1

[2] Annual smoking-attributable mortality, years of potential life lost, and economic cost-united state 1995-1999, MMWR morb mortal wkly Rep,51(2002), pp.300-303

[3] A. Jemal, M. M. center, C. Desantis EM Ward, Global patterns of cancer incidence and mortality rates and trends, cancer Epidemiol Biomarkers prev, 19(2010) pp. 1893-1907

[4] D. Hoffmann, I. Hoffmann, K. El-Bayoumy, The less harmful cigarette: a controversial issue. A tribute to Ernst L Wynder, chem. Res Toxicol, 14(2001), pp.767-790

[5] A. Charlton, Medicinal use of tobacco in history, J R Med, 97 (2004), pp. 292-296

[6] P. Gordia, H. Lchikawa, N. Malani, G. Sethi, B. B. Aggarwal, From ancient medicine to modern medicine:ayurvedic concepts of health and their role in inflammation and cancer, J Soc Integr Oncol, 5 (2007), pp. 25-27

[7] J. Handelsman, J. Conway, M. Boylan, R. Turtle, Testicular function in potential sperm ldonors: normal ranges and the effect of smoking and varicocele, Int J Androl, 7(1984), pp. 369-382

[8] Atangana, A., Alkahtani, B. S. T.: Analysis of the Keller–Segel model with a fractional derivative without singular kernel. Entropy 17(6), 4439–4453 (2015)

[9] Alsaedi, A., Nieto, J. J., Venktesh, V.: Fractional electrical circuits. Adv. Mech. Eng. 7(12), 1687814015618127 (2015)

[10] Caputo, M., Fabrizio, M.: A new definition of fractional derivative without singular kernel. Prog. Fract. Differ. Appl. 1(2), 1–13 (2015)

[11] Losada, J., Nieto, J. J.: Properties of a new fractional derivative without singular kernel. Prog. Fract. Differ. Appl. 1(2), 87–92 (2015)

[12] Tateishi, A. A., Ribeiro, H. V., Lenzi, E. K.: The role of fractional time-derivative operators on anomalous diffusion. Front. Phys. 5, 52 (2017)

[13] Kumar, D., Singh, J., Al Qurashi, M., Baleanu, D.: Analysis of logistic equation pertaining to a new fractional derivative with non-singular kernel. Adv. Mech. Eng. 9(2), 1687814017690069 (2017)

[14] Owolabi, K. M., Atangana, A.: Analysis and application of new fractional Adams–Bashforth scheme with Caputo–Fabrizio derivative. Chaos Solitons Fractals 105, 111–119 (2017)

[15] Kumar, D., Tchier, F., Singh, J., Baleanu, D.: An efficient computational technique for fractal vehicular traffic flow. Entropy 20, 259 (2018)

[16] Singh, J., Kumar, D., Baleanu, D.: New aspects of fractional Biswas–Milovic model with Mittag–Leffler law. Math. Model. Nat. Phenom. 14(3), 303 (2019)

[17] Diethelm, K.: The Analysis of Fractional Differential Equations: An Application-Oriented Exposition Using Differential Operators of Caputo Type. Springer, Berlin (2010)

[18] Atangana, A.: Blind in a commutative world: simple illustrations with functions and chaotic attractors. Chaos Solitons Fractals 114, 347–363 (2018)

[19] Tateishi, A. A., Ribeiro, H. V., Lenzi, E. K.: The role of fractional time-derivative operators on anomalous diffusion. Front. Phys. 5, 52 (2017)

[20] H. Bulut, Baskonus, H. M, Belgacem, FBM: The analytic solutions of some fractional ordinary differential equation by Sumudu transform method, Abstr.Appl.Anal.2013.Article ID203875(2013)

[21] Kilbas, A. A., Srivastava, H. M., Trujillo, J. J.: Theory and Applications of fractional differential equations. Elsevier, Amsterdam (2006)

[22] Atangana, A., Alkahatani BT: Analysis of non-homogeneous heat model with new trends of derivative with fractional order, Chaos Solitons fractals 89, 566-571 (2016)

[23] Huang, L. L., Baleanue, D., Wu Zeng, S. D.: A new application of fractional logistic map Rom J phys. 61, 1172-1179 (2016)

[24] Zaman, G., Qualitative behavior of giving up smoking model, Bull. Malays Math. Soc. 34,403-415(2011)

[25] Ertuk vs Zaman G., mamoni S., A numeric analytic method for approximating a giving up smoking model containing fractional derivatives (2011)

2

Hybrid Feature Selection Techniques Utilizing Soft Computing Methods for Classifying Microarray Cancer Data

Rabia Musheer Aziz[1], Amol Avinash Joshi[2,*], Kartik Kumar[3], and Abdu Hamid Gaani[4]

[1,2,3]Department of SASL (Mathematics)
[3]Department of Computer Seience, VIT Bhopal University, Bhopal- Indore Highway, Kothrikalan, Sehore, -466116 (M.P.) India
[4]Saudi Electronic University

Abstract

Recently, various soft computing techniques have been used to extract k information from big data. A standardized format for evaluating the expression levels of thousands of genes is made available by DNA microarray technology. The cancers of several anatomical regions can be identified with the help of patterns developed by gene expressions in microarray technology. As the microarray data is too huge to process due to the curse of dimensionality problem. Therefore, in this paper, a hybrid machine learning framework using soft computing techniques for feature selection is designed and executed to eliminate unnecessary genes and identify important genes for the identification of cancer. In the first stage, the genes or the features are taken out with the aid of the higher-order Independent Component Analysis (ICA) technique. Then in the second level, Genetic Bee Colony (GBC) optimization techniques are utilized in this work for selecting the best genes or features before proceeding to the classification process. For the comparison purpose, three other optimization techniques are considered in this work, which are Particle Swarm Optimization (PSO), Artificial Bee Colony (ABC), and Genetic Algorithm (GA). After the selection of important genes, the two most popular classifiers Naïve Bayes (NB) and Support Vector Machine (SVM)) are trained with selected genes and at last, find classification accuracy of test data. The experimental results with five benchmark microarray datasets of cancer, prove that GBC is a more efficient approach to improve the classification performance with ICA for both the classifiers.

Keyword: Genetic Bee Colony (GBC), Independent Component Analysis (ICA), Genetic Algorithm (GA) particle swarm optimization (PSO), Support Vector Machine (SVM) and Naïve Bayes (NB).

2.1 Introduction

A cancer is zero but the unnatural growth of cells in the affected area, and owns the ability to distribute it to other vital areas of the physical structure [1–3]. The cancer classification based on gene expression profiles has provided better transparency for the possible treatment strategies. Recently, because of the increase in data generation and storage, various big data applications gained attention, which also increased interest in the application of the same, to a wide range of biological problems [1, 4]. The identification of genes plays an important role in detecting cancer diseases and has an essential impact on the microarray cancer prediction [5, 6]. Therefore, dimension reduction followed by classification acts as the major process for further analysis. Currently, due to time complexity in the current data technology, the big data is increased. Cancer data of microarray data is also a type of big, big data analytics that uses machine learning and soft computing techniques for the extraction of knowledge and other insights from big data. A huge number of datasets creates complications for the researchers who are trying to identify useful information through the applications of statistical and bioinformatics tools. This research has moved to a new age of molecular classification [7–9]. For the prediction of cancer disease, the microarray data classification technique is utilized widely [10]. To monitor genome-wide expression, one of the vital tools that many biologists rely on is microarray technology. The gene expression data classification is an important task, due to the presence of a huge no of genes related to samples. Mostly, many genes will not be more informative either they may be irrelevant or redundant. For this purpose the number of genes needs to be reduced for improving the accuracy of the classification task. So, classification with this huge amount of data is difficult as it increases computational cost thereby degrading the performance of the classifier. Therefore, for such asymmetric data, it is very difficult to utilize the traditional classifiers and therefore the analysis of microarray data, a dimensional reduction is highly required [11]. The Dimensionality reduction approaches are used to find relevant and specific information in a huge number of datasets, which are known as feature selection and feature extraction [12]. Classification of diverse types of predominant gene expression in enumerating genes for biological questions has given beneficial information for identifying and discovering drugs [9, 13].Recently, Gene expression analysis was used to find out the risk analysis of cancer by using data mining with soft computing techniques. Some of the researchers used multiple soft computing paradigms, the statistical characterization for feature selection, and the cancer classification of microarray. Various supervised and unsupervised techniques with soft computing methods are useful tools that generally yield high accuracy in the classification and prediction process. In [14] author improved the performance of the Artificial Bee Colony by using a Genetic algorithm for SVM classification. The proposed algorithm has been implemented with different microarray datasets for the selection of informative genes for cancer classification. The paper showed the high classification accuracy with a low average number of the genes selected. For the diagnosis and survival prediction of cancer, most the researchers utilize an intelligent technique for gene selection and classify it with different classifiers. [15] Presented a review on machine learning-based various genes selection algorithms. The paper found that gene selection chooses the relevant and important genes and removes the irrelevant genes for improving the classification accuracy of different machine learning classifiers. Now a day to reduce the computational time and to take the benefits of the different dimension reduction methods combination of different methods is used as a hybrid approach [16, 17]. Hybrid approaches combine different feature selection and extraction methods to reduce the dimension of the data. Different researchers applied a different combination of the algorithm according to the requirement of different data sets. [18] Proposed an effective ensemble approach for three benchmark data sets leukemia dataset, colon dataset, and breast cancer dataset with KNN classifier, which helps to increase the performance of the classification. In [19] author combined PSO with Artificial Bee Colon for SVM classification of nine cancer datasets and found better results compared to individual PSO and ABC. [20] has surveyed the feature selection methods for gene expression. It focuses on the filter, wrapper, and embedded methods and found a hybrid approach based on filter and wrapper gives better results of data in machine learning problems or pattern reorganization applications. Some authors designed a classification framework and applied it to classify cancer gene expression profiles with various hybrid gene selection algorithms based on different nature-inspired metaheuristic techniques [16]. [21] Proposed classification model that estimates the reduced gene sets quality with a hybrid approach based on optimum path forest and bat algorithm. [22] Proposed a trending PSO variant as CSO to resolve the advanced feature selection problem. The work not only selected very few features but also reduced computational cost by using the collection of new techniques that produced a good performance in classification. A comparison result expresses that the proposed hybrid approach has been successfully applied and excels with other existing methods in terms of accuracy. In [23] author selected the important feature by using robust principal component analysis and classified tumors with SVM. Evaluated the performance of the proposed approach with seven data sets and found the proposed approach was effective for classification. Author [24] used the ICA extraction method with stepwise regression for feature selection on five benchmark microarray datasets and proposed approach demonstrated to improving the classification performance of NB classifier [25–28]. In [29] proposed a three-level ensemble approach for cancer prediction and classification. For gene selection firstly ensemble Fisher ratio and T-test after that optimize the gene with PSO-dICA method then classify five cancer microarray data sets and find satisfactory results compare to other published results. In [30] selected the feature subset by using ICA that extracted essential information and at the same time separated the noise as extracted features were an independent component. The results of the data sets have shown the effectiveness and improvement of the proposed approach. In [31, 32] the author used nature-inspired metaheuristic techniques-based wrapper methods with ICA and found ICA increase the classification accuracy of a different classification algorithm for cancer microarray profile. Nature-inspired metaheuristic methods-based hybrid approaches have successfully resolved the problem of many fields and increased the classification accuracy of various classifiers. These nature-inspired algorithms are not applied directly to gene expression microarray data due to the high dimensionality that increases the computational task. Many researchers take the advantage of nature-inspired metaheuristic algorithms through hybrid models with numerous combinations of machine learning approaches for gene selection of microarray data [33–35]. In this study, a hybrid heuristic dimensionality reduction technique is proposed to classify different cancer types from gene expression microarray data and identify significant features by exploiting Genetic Bee Colony (GBC), as an optimization technique for finding the best gene subset from ICA extracted feature sets. Finally, applied the proposed approach with NB and SVM classifiers and evaluate the classification accuracy of five benchmark microarray data sets. Figure 2.1, Show the framework of the proposed algorithm.

2.2 Proposed Framework

2.2.1 Genes Extraction by ICA

The ICA method helps in obtaining hidden features from multi-dimensional information, by decomposing multi variate indications into independent non-gaussian sections for the components, to be statistically independent. ICA finds a correlation between data by decorrelating the data, by exploiting or diminishing the distinct information. ICA assumes all genes are X as an amalgam of independent components S. If A signifies the opposite matrix of a weighted matrix W, and columns of A, characterize the source featurevectors of comment X [36].

$$S = W \times X, \quad X = A \times S$$

ICA has been extensively used for biological information, recognitions and also applicable to another domain

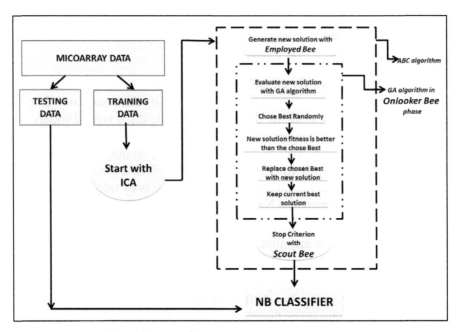

Figure 2.1 Overall flow of the proposed framework.

2.2.2 Genetic Bee Colony (GBC) Algorithm

The genetic Bee (GBC) technique is a combination of ABC and the Genetic Algorithm (GA) algorithm [39]. GBC is a novel technique; the purpose of GBC techniques is to find the best subsets of a gene for improving the accuracy of the different classification algorithms. To achieve the highest classification accuracy with the ABC algorithm, an appropriate equilibrium between exploitation and exploration is essential [40]. The exploration process of the ABC algorithm for finding a new solution compare to the other nature expired algorithm is good, but the exploitation process required more computational time to converge to the optimal solution [31, 41]. Therefore, to reduce the computational time of the exploitation process of ABC, the GBC algorithm used GA. The exploitation process of the GA algorithm is good due to the use of crossover and mutation operations, but its ability for searching for a new solution in the optimization search space is not good as compared to the ABC algorithm [42, 43]. That is why the GA obtained the local optima too quickly and suffers from the pre-convergence problem which is the main issue with the GA algorithm. Therefore, to maintain the balance between the exploitation and exploration process, the GBC algorithm uses a combination of ABC and GA algorithms [14].

2.3 Used Classifier

2.3.1 Naïve Bayes Classifier (NBC)

In this paper, NBC is used for classifying cancer samples into two different classes, it is widely used by many researchers, to classify the objects into 2 or more classes by means of using the Bayes theorem [44, 45]. It is used widely, when the input variable is continuous and independent, then the parameters are estimated by the Bayes rule, so that the probability of output variable is exactly predicted [46, 47]. If E_1, E_2,...,E_n. are the selected genes from any sample of H, Naïve Bayes classifier classified the samples by using the below formula with Bayes theorem as a Naïve Bayes classifier [37, 48]:

$$H' = \arg \max_{H \in \omega} P(H) \prod_{i=1}^{n} f(E_i|H)$$

Because features of microarray are continuous, so for the calculation of class-conditional probability i.e. $f(.|H)$, probability density function with nonparametric kernel density estimation method, each attribute is used and $P(H)$ is the prior probability of the particular class [49, 50].

2.3.2 Support Vector Machine (SVM) Classifier

SVM is the most popular classifier for microarray data as SVM finds insightful hyperplanes, that optimally segment tuples of a course of study from some other course of study. Hyperplanes are initiated from the gross profits and support vectors calculated from the directions defining the hyperplane. SVM generates result boundaries between positive and negative groups by selecting the utmost pertinent instances intricated in the decision procedure. When data are linear, and discrete, the hyperplane construction is constantly conceivable. SVM uses kernels, where non linear maps into high dimension feature space, for separating the hyperplanes found earlier. SVM kernel includes polynomial kernel, linear kernel, String Kernels, Radial basis function (RBF), Gaussian kernel, and Sigmoid kernel. In the case of nonlinear distinguishable data, the kernal function maps the data in high dimensional space with the help of the below optimization technique used in the SVM algorithm [51–53]:

$$\min \text{ imize} \frac{1}{2} \| W^T \|^2 + C \sum_{i=1}^{n} \xi_i$$
$$\text{subject to} : Y_i (W^T K(X_i) + b) \geq 1 - \xi_i$$
$$\xi_i \geq 0$$

where K is different kernal function which is used according to the requirement of different data sets and is slack variable that helps to find the maximum margin between two classes.

2.4 Experimental Setups

The extensive experiment conducted with five datasets of cancer microarray. Table 2.1 displays characteristics of five microarray data set that used in this experiment.

Table 2.1 Summary of five benchmark microarray gene expression datasets.

Data set	Colon cancer (Alon et al.) [54]	Acute leukemia (Golub et al.) [55],	Prostate tumor (Dinesh Singh et al.) [56]	High-grade Glioma (C.L. Nutt et al.) [57],	Lung cancer II (Gorden et al.) [58]
No. of genes	2000	7129	12600	12625	12533
No. of samples	62	72	102	50	181

To check the performance of the proposed approach, firstly we need to train the classification model with training datasets and it should predict the class of the test data set. Before constructing the classification model, a reduction in the number of genes should be carried out with the help of the proposed algorithm. Tables 2.2–2.6 depicted the Classification accuracy

Table 2.2 Classification accuracy of NB and SVM classifiers with best selected gene sets by using different feature selection technique for colon cancer data

S. No.	Classifier	Method	Mean accuracy	Variance
1.		ICA+SVM	78.17	0.069
2.		ICA+GA+SVM	78.25	0.054
3.	SVM	ICA+PSO+SVM	81.16	0.048
4.		ICA+ABC+SVM	85.45	0.031
5.		ICA+GBC+SVM	92.07	0.021
1.		ICA+PSO+NB	80.81	0.076
2.		ICA+GA+NB	85.16	0.061
3.	NB	ICA+ABC+NB	87.31	0.029
4.		ICA+GBC+NB	91.12	0.014

Table 2.3 Classification accuracy of NB and SVM classifiers with best selected gene sets by using different feature selection technique for colon cancer data for Acute leukemia data

S. No.	Classifier	Method	Mean accuracy	Variance
1.		ICA+SVM	80.21	0.073
2.		ICA+GA+SVM	78.77	0.036
3.	SVM	ICA+PSO+SVM	86.43	0.027
4.		ICA+ABC+SVM	92.04	0.027
5.		ICA+GBC+SVM	96.30	0.011
1.		ICA+PSO+NB	86.21	0.063
2.		ICA+GA+NB	92.77	0.041
3.	NB	ICA+ABC+NB	94.62	0.032
4.		ICA+GBC+NB	96.55	0.021

Table 2.4 Classification accuracy of NB and SVM classifiers with best selected gene sets by using different feature selection technique for colon cancer data for Prostate tumor data.

S. No.	Classifier	Method	Mean accuracy	Variance
1.		ICA+SVM	73.43	0.102
2.		ICA+GA+SVM	76.43	0.089
3.	SVM	ICA+PSO+SVM	78.48	0.072
4.		ICA+ABC+SVM	81.72	0.066
5.		ICA+GBC+SVM	83.52	0.033
1.		ICA+PSO+NB	79.23	0.092
2.		ICA+GA+NB	81.12	0.073
3.	NB	ICA+ABC+NB	84.82	0.043
4.		ICA+GBC+NB	83.94	0.028

Table 2.5 Classification accuracy of NB and SVM classifiers with best selected gene sets by using different feature selection technique for colon cancer data for High-grade Glioma data.

S. No.	Classifier	Method	Mean accuracy	Variance
1.		ICA+SVM	74.23	0.069
2.		ICA+GA+SVM	75672	0.047
3.	SVM	ICA+PSO+SVM	78.51	0.039
4.		ICA+ABC+SVM	84.12	0.027
5.		ICA+GBC+SVM	91.71	0.019
1.		ICA+GA+NB	75.20	0.052
2.		ICA+PSO+NB	83.97	0.047
3.	NB	ICA+ABC+NB	87.44	0.035
4.		ICA+GBC+NB	89.36	0.023

Table 2.6 Classification accuracy of NB and SVM classifiers with best selected gene sets by using different feature selection technique for colon cancer data for Lung cancer II data.

S. No.	Classifier	Method	Mean accuracy	Variance
1.		ICA+SVM	71.21	0.084
2.		ICA+GA+SVM	75.77	0.073
3.	SVM	ICA+PSO+SVM	85.28	0.076
4.		ICA+ABC+SVM	87.31	0.052
5.		ICA+GBC+SVM	92.63	0.021
1.		ICA+GA+NB	86.52	0.067
2.		ICA+PSO+NB	83.76	0.061
3.	NB	ICA+ABC+NB	85.23	0.054
4.		ICA+GBC+NB	87.48	0.021

of NB and SVM classifiers, with best selected gene sets, by using different feature selection techniques for all five cancer datasets. All data sets are classified with the help of a LOOCV cross validation method, performance Index, specificity, and sensitivity are used, and the values and mean accuracy are computed by taking the average of sensitivity and specificity exhibited in Tables 2.2 to 2.6. Firstly, extract the features using ICA from the training data set, and extracted features are optimized using the GBC algorithm, at the end, the classification accuracy of both the classifiers is evaluated with the best subset of features.

2.5 Experimental Result

In this study of the classification of microarray cancer data, dimensionality was reduced with a hybrid approach using a combination of unsupervised algorithm ICA and new improved optimization technique GBC. The proposed algorithm (ICA+GBC) achieved 92.07% and 91.12% for colon cancer data,

96.30% and 96.55% for acute leukemia, 83.52% and 83.94% for prostate cancer, 91.71% and 89.36% for high grade glioma and 92.63% and 87.48% for lung cancer II data classification accuracy respectively with both the classifiers. The performance evaluation of all the methods is tabulated below, comparing methods, ICA+GBC+SVM and ICA+GBC+NB were found to perform superior to others, it is of essence to note that even the NB classifier is less popular compared to the SVM classifier for microarray data, yet it presented a higher average accuracy for two data sets. Therefore, it is clear from the results ICA+GBC method provided a reduced number of genes that gives more accurate, stable results for both the classifiers.

Figures 2.2 and 2.3 shows the average error rate graph of NB and SVM classifier with five hybrid features selection method of five microarray cancer data sets respectively. These figures show the significance of the hybrid method compared to a single method for microarray data for SVM and NB classifiers. It clearly shows in the figure that the classification accuracy increases when we applied the hybrid method instead of one method and the (ICA+GBC), the proposed algorithm provided the best accuracy rate compared to the other hybrid method with SVM and NB classifier. Since, irrelevant, noisy, and redundant genes create the problem for building the classification model and increase the problem of overfitting. Two resolved this problem in this research ICA feature extraction method is used with different

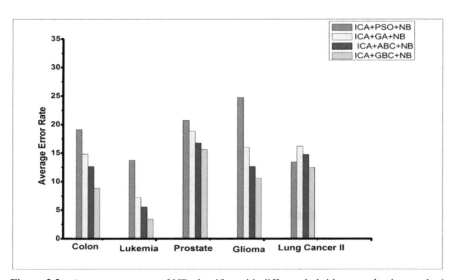

Figure 2.2 Average error rate of NB classifier with different hybrid gene selection method for the five microarray cancer datasets.

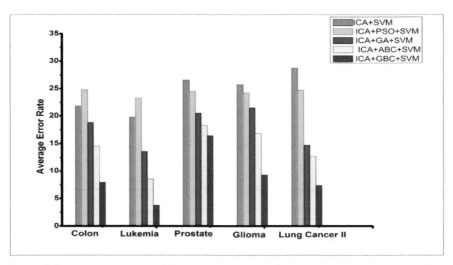

Figure 2.3 Average error rate of SVM classifier with different hybrid gene selection method for the five cancer microarray datasets.

optimization methods for removing irrelevant and unimportant genes for SVM and NB classification. For selecting genes from the ICA features vector used the GBC wrapper approach as a hybrid algorithm for the classification of microarray data, so we selected a different number of genes from ICA feature vectors with the different optimization methods. Find the best number of features by evaluating the classification accuracy of the classifier by wrapper approach with classifiers and find the mean accuracy and variance of the proposed algorithm. The above step is applied and repeated for all data sets for NB and SVM classification until all-important genes are received. Figures 2.4-2.8 indicate the different number of genes with the classification performance of both classifiers for all five data sets with the proposed algorithm.

In Figures 2.4-2.8, the peak of the graph depicts the sufficient number of genes to categorize several samples into their respective class. The size of extracted genes was decreased by GBC with classifiers as a wrapper approach and repeated the process till got the best subset. Results express that the proposed method selected the best subset of genes that provided a high classification of both the classifier accuracy compared to other published popular methods. Therefore, the result concludes that the proposed approach reduced the dimensionality of microarray and find the best subset of the gene that increases the classification accuracy of the classifiers.

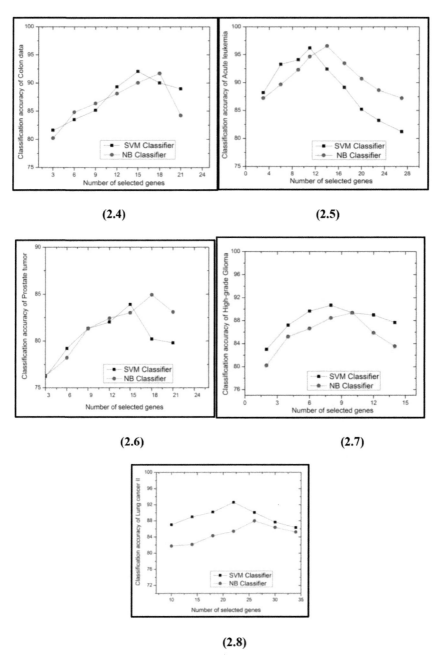

Figure 2.4-2.8 Classification accuracy of NB and SVM classifiers with different number of genes with proposed approach (ICA+GBC) for above define five data sets.

2.6 Conclusion

The proposed approach (ICA+GBC), selected relevant, important, and informative features that increase the classification accuracy of NB and SVM classifiers. Therefore, the proposed approach reduced the dimensionality and resolved the problem of the building of the classification model, which is a challenging issue in high dimensional gene expression data Several approaches have been proposed for the enhancement of the technology, towards the prediction and detection of ailments derived from samples, dimensionality reduction has proven to solve these challenges, yet more investigations need to be carried out. Recently, hybrid approaches have been used in recent time for the classification of gene expression data. To ascertain these approaches, these experiments carried out a dimensionality reduction with a hybrid approach using soft computing techniques for five benchmarks microarray datasets and tested their evaluation performance on SVM and NB classifiers. Experimental results found that ICA+GBC performed better than the other ICA-based hybrid approaches for both the classifiers.

References

[1] Makrariya, Akshara, and Neeru Adlakha. "Quantitative study of thermal disturbances due to nonuniformly perfused tumors in peripheral regions of women's breast." Cancer informatics 16 (2017): 1176935117700894.

[2] Lu, Y. and J. Han, Cancer classification using gene expression data. Information Systems, 2003. 28(4): p. 243-268.

[3] Makrariya, Akshara, and Neeru Adlakha. "Thermal stress due to tumor in periphery of human breast." J Biosci Bioeng 2 (2015): 50-59.

[4] Xu, Y., et al., Artificial neural networks and gene filtering distinguish between global gene expression profiles of Barrett's esophagus and esophageal cancer. Cancer Research, 2002. 62(12): p. 3493-3497.

[5] Makrariya, A., Adlakha, N., & Shandilya, S. K. (2021). 3D Spherical—Thermal Model of Female Breast in Stages of Its Development and Different Environmental Conditions. In Mathematical Modeling, Computational Intelligence Techniques and Renewable Energy: Proceedings of the First International Conference, MMCITRE 2020 (pp. 217-227). Springer Singapore.

[6] Elek, J., K. Park, and R. Narayanan, Microarray-based expression profiling in prostate tumors. In vivo (Athens, Greece), 1999. 14(1): p. 173-182.

[7] Selvakumar, K., et al., Intelligent temporal classification and fuzzy rough set-based feature selection algorithm for intrusion detection system in WSNs. Information Sciences, 2019. 497: p. 77-90.

[8] Santhakumar, D. and S. Logeswari, Efficient attribute selection technique for leukaemia prediction using microarray gene data. Soft Computing, 2020: p. 1-10.

[9] Hoque, N., M. Singh, and D.K. Bhattacharyya, EFS-MI: an ensemble feature selection method for classification. Complex & Intelligent Systems, 2018. 4(2): p. 105-118.

[10] Musheer, R.A., C. Verma, and N. Srivastava, Novel machine learning approach for classification of high-dimensional microarray data. Soft Computing, 2019. 23(24): p. 13409-13421.

[11] Aziz, R., C. Verma, and N. Srivastava, Dimension reduction methods for microarray data: a review. 2017.

[12] Makrariya, Akshara, and Neeru Adlakha. "Thermal Patterns in Peripheral Regions of Breast during Different Stages of Development." Biology and Medicine 9.1 (2017): 1.

[13] Aziz, R., C. Verma, and N. Srivastava, Artificial neural network classification of high dimensional data with novel optimization approach of dimension reduction. Annals of Data Science, 2018. 5(4): p. 615-635.

[14] Alshamlan, H.M., G.H. Badr, and Y.A. Alohali, Genetic bee colony (GBC) algorithm: a new gene selection method for microarray cancer classification. Computational biology and chemistry, 2015. 56: p. 49-60.

[15] Makrariya, A., & Pardasani, K. R. (2018). Finite Element Model To Study The Thermal Effect Of Cyst And Malignant Tumor In Women's Breast During Menstrual Cycle Under Cold Environment. Advances and Applications in Mathematical Sciences, 18(1), 29-43.

[16] Mafarja, M., et al., Efficient hybrid nature-inspired binary optimizers for feature selection. Cognitive Computation, 2020. 12(1): p. 150-175.

[17] Venkatesh, B. and J. Anuradha, A review of feature selection and its methods. Cybernetics and Information Technologies, 2019. 19(1): p. 3-26.

[18] Tarek, S., R. Abd Elwahab, and M. Shoman, Gene expression based cancer classification. Egyptian Informatics Journal, 2017. 18(3): p. 151-159.

[19] Gao, L., M. Ye, and C. Wu, Cancer classification based on support vector machine optimized by particle swarm optimization and artificial bee colony. Molecules, 2017. 22(12): p. 2086.

[20] Chandrashekar, G. and F. Sahin, A survey on feature selection methods. Computers & Electrical Engineering, 2014. 40(1): p. 16-28.

[21] Rodrigues, D., et al., A wrapper approach for feature selection based on bat algorithm and optimum-path forest. Expert Systems with Applications, 2014. 41(5): p. 2250-2258.

[22] Gu, S., R. Cheng, and Y. Jin, Feature selection for high-dimensional classification using a competitive swarm optimizer. Soft Computing, 2018. 22(3): p. 811-822.

[23] Liu, J.-X., et al., RPCA-based tumor classification using gene expression data. IEEE/ACM Transactions on Computational Biology and Bioinformatics, 2014. 12(4): p. 964-970.

[24] Fan, L., K.-L. Poh, and P.J.E.S.w.A. Zhou, A sequential feature extraction approach for naïve bayes classification of microarray data. 2009. 36(6): p. 9919-9923.

[25] Zheng, C.-H., D.-S. Huang, and L. Shang, Feature selection in independent component subspace for microarray data classification. Neurocomputing, 2006. 69(16): p. 2407-2410.

[26] Zheng, C.-H., et al., Gene expression data classification using consensus independent component analysis. Genomics, proteomics & bioinformatics, 2008. 6(2): p. 74-82.

[27] Rabia, A., S. Namita, and K.V. Chandan, t-Independent Component Analysis For SVM Classification Of DNA- Microarray Data. International Journal of Bioinformatics Research, 2015. 6(1): p. 305-312.

[28] Rabia, A., S. Namita, and K.V. Chandan, A Weighted-SNR Feature Selection from Independent Component Subspace for NB Classification of Microarray Data. International Journal of Advanced Biotechnology and Research, 2015. 6(2): p. 245-255.

[29] Mollaee, M., M.H.J.B. Moattar, and B. Engineering, A novel feature extraction approach based on ensemble feature selection and modified discriminant independent component analysis for microarray data classification. 2016. 36(3): p. 521-529.

[30] Mahdavi, K., J. Labarta, and J. Gimenez. Unsupervised Feature Selection for Noisy Data. in International Conference on Advanced Data Mining and Applications. 2019. Springer.

[31] Aziz, R., et al., Artificial neural network classification of microarray data using new hybrid gene selection method. International Journal of Data Mining and Bioinformatics, 2017. 17(1): p. 42-65.

[32] Aziz, R., C. Verma, and N. Srivastava, A Novel Approach for Dimension Reduction of Microarray. Computational Biology and Chemistry, 2017.

[33] Gheyas, I.A. and L.S. Smith, Feature subset selection in large dimensionality domains. Pattern recognition, 2010. 43(1): p. 5-13.

[34] Unler, A., A. Murat, and R.B. Chinnam, mr 2 PSO: a maximum relevance minimum redundancy feature selection method based on swarm intelligence for support vector machine classification. Information Sciences, 2011. 181(20): p. 4625-4641.

[35] Tabakhi, S., P. Moradi, and F. Akhlaghian, An unsupervised feature selection algorithm based on ant colony optimization. Engineering Applications of Artificial Intelligence, 2014. 32: p. 112-123.

[36] Hyvarinen, A. and J. Karhunen, Oja., E. Independent component analysis. John Wiley&Sonr, 2001.

[37] Aziz, R., C. Verma, and N. Srivastava, A Fuzzy Based Feature Selection from Independent Component Subspace for Machine learning Classification of Microarray Data. Genomics Data, 2016.

[38] Hsu, C.-C., M.-C. Chen, and L.-S. Chen, Integrating independent component analysis and support vector machine for multivariate process monitoring. Computers & Industrial Engineering, 2010. 59(1): p. 145-156.

[39] Karaboga, D., An idea based on honey bee swarm for numerical optimization. 2005, Technical report-tr06, Erciyes university, engineering faculty, computer engineering department.

[40] Garro, B.A., K. Rodríguez, and R.A. Vázquez, Classification of DNA microarrays using artificial neural networks and ABC algorithm. Applied Soft Computing, 2016. 38: p. 548-560.

[41] Garro, B.A., K. Rodríguez, and R.A. Vázquez, Classification of DNA microarrays using artificial neural networks and ABC algorithm. Applied Soft Computing, 2015.

[42] Kiran, M. S., et al., A novel hybrid approach based on particle swarm optimization and ant colony algorithm to forecast energy demand of Turkey. Energy conversion and management, 2012. 53(1): p. 75-83.

[43] Jatoth, R.K. and A. Rajasekhar. Speed control of pmsm by hybrid genetic artificial bee colony algorithm. in Communication Control and Computing Technologies (ICCCCT), 2010 IEEE International Conference on. 2010. IEEE.

[44] Friedman, N., D. Geiger, and M. Goldszmidt, Bayesian network classifiers. Machine learning, 1997. 29(2-3): p. 131-163.

[45] Hall, M., A decision tree-based attribute weighting filter for naive Bayes. Knowledge-Based Systems, 2007. 20(2): p. 120-126.

[46] Chen, J., et al., Feature selection for text classification with Naïve Bayes. Expert Systems with Applications, 2009. 36(3): p. 5432-5435.

[47] Sandberg, R., et al., Capturing whole-genome characteristics in short sequences using a naive Bayesian classifier. Genome research, 2001. 11(8): p. 1404-1409.

[48] Fan, L., K.-L. Poh, and P. Zhou, A sequential feature extraction approach for naïve bayes classification of microarray data. Expert Systems with Applications, 2009. 36(6): p. 9919-9923.

[49] Fan, L., K.-L. Poh, and P. Zhou, Partition-conditional ICA for Bayesian classification of microarray data. Expert Systems with Applications, 2010. 37(12): p. 8188-8192.

[50] De Campos, L.M., et al. Bayesian networks classifiers for gene-expression data. in Intelligent Systems Design and Applications (ISDA), 2011 11th International Conference on. 2011. IEEE.

[51] Cortes, C. and V. Vapnik, Support-vector networks. Machine learning, (3)1995. 20(3): p. 273-297.

[52] Vapnik, V., Statistical learning theory. 1998. 1998, Wiley, New York.

[53] Mukherjee, S. and V. Vapnik, Support vector method for multivariate density estimation. Center for Biological and Computational Learning. Department of Brain and Cognitive Sciences, MIT. CBCL, 1999. 170.

[54] Alon, U., et al., Broad patterns of gene expression revealed by clustering analysis of tumor and normal colon tissues probed by oligonucleotide arrays. Proceedings of the National Academy of Sciences, 1999. 96(12): p. 6745-6750.

[55] Golub, T.R., et al., Molecular classification of cancer: class discovery and class prediction by gene expression monitoring. science, 1999. 286(5439): p. 531-537.

[56] Singh, D., et al., Gene expression correlates of clinical prostate cancer behavior. Cancer cell, 2002. 1(12): p. 203-209.

[57] Gordon, G.J., et al., Translation of microarray data into clinically relevant cancer diagnostic tests using gene expression ratios in lung cancer and mesothelioma. Cancer research, 2002. 62(17): p. 4963-4967.

[58] Nutt, C.L., et al., Gene expression-based classification of malignant gliomas correlates better with survival than histological classification. Cancer research, 2003. 63(7): p. 1602-1607.

3

Finite Element Technique to Explicate Calcium Diffusion in Alzheimer's Disease

Devanshi D. Dave and Brajesh Kumar Jha

Pandit Deendayal Energy University
E-mail: ddave1822@gmail.com

Abstract

Calcium is ubiquitously known as one of the most important second messengers. Being divalent in nature, calcium easily binds with certain buffers and maintains the cytosolic calcium concentration level. The free calcium in the cell is mainly regulated by two phenomena; one is entry through certain calcium channels and secondly, intracellular calcium is maintained by calcium stores such as the endoplasmic reticulum. Hence, the alteration in the cell calcium results in fatal neurodegenerative diseases. Alzheimer's disease is a progressing neurodegenerative disorder that is the result of a high cytosolic calcium level. In this chapter, a two-dimensional mathematical model is developed to study the impact of buffers, voltage-gated calcium channels, and endoplasmic reticulum on the calcium concentration distribution in Alzheimer's cells. The irregular-shaped geometry of the cell has been considered over here. All necessary physiological parameters are taken into consideration to get the results which are obtained by finite element technique and simulated using MATLAB. The results obtained clearly shows the significant changes taking place in normal and Alzheimer's affected neuronal cells.

Keywords: Calcium, buffers, voltage gated calcium channel, endoplasmic reticulum, Alzheimer's disease, Finite element technique.

Mathematics Subject Classification (2010) 35Q92, 62P10, 92C35, 74S05

3.1 Introduction

Alzheimer's disease is the most common form of dementia affecting millions of people worldwide. Being sporadic in nature, no perfect cause for Alzheimer's to prevail is known to date [1]. Several studies and researchers have failed in predicting the exact pathophysiology of Alzheimer's whereas studies are still going on in knowing the pros and cons of Alzheimer's disease. Mainly people having the age group of 60 and above are highly affected by Alzheimer's symptoms [2]. Moreover, a study shows that over and above the age group, Alzheimer's disease is likely to prevail more in females as compared to males. The risk factor for females is 2-3 times more than for males and the main reason for it may be the deficiency of estrogen in the brain due to menopause [3]. Alzheimer's disease is progressive and irreversible which leads to cognitive, be- behavioral, and memorial impairments. Due to the idiopathic nature of Alzheimer's disease, one cannot pinpoint the only actual cause of it to prevail, and hence the mechanism underlying the causes is not known to date. The physiological changes which are known as pathological hallmarks are neurofibrillary tangles and amyloid-beta plaques [4]–[6], whereas the symptoms of Alzheimer's are loss of memory and linguistic skills, behavioral impairments, etc. [4]–[7].

The brain tissue of Alzheimer's patients shows evidence of changes in the regular calcium homeostasis [8]. The alteration in the cellular calcium results in neurodegenerativity. The cell calcium in the human brain plays an important role in maintaining a plethora of functions [9]. Calcium level in the cell is maintained by calcium-binding proteins, several intracellular entities like endoplasmic reticulum and mitochondria and certain exchangers and calcium channels like sodium-calcium exchanger, voltage-gated calcium channels, and some of the receptors like ryanodine and IP3 receptors [9]. Thus, all the above-mentioned parameters work as a syncytium and maintain the calcium level in the cell. Many the researchers/studies have revealed that there is a linkage between calciumopathy and Alzheimer's disease [4], [10], [11]. Long back in the 1970's, Khachaturian laid the foundation, linking cytosolic calcium and Alzheimer's disease [12]. Since then, the era linking calcium and Alzheimer's began. It has been found that a higher level of calcium results in the pathology of Alzheimer's disease. Moreover, the hallmark symptoms of Alzheimer's. i.e., neurofibrillary tangles and amyloid beta plaques have been found to take place due to alterations in calcium homeostasis [5], [10], [13], [14]. Thus, calcium and Alzheimer's are closely related. It has been found that mostly calcium is found in the free state at the cytosolic level where it bounds with the calcium binding proteins present in the peripheral region of the cell [15]. The buffer by forming a calcium bound buffer maintains a higher level of calcium in the cell and regulates many physiological functions. On the other hand, the flux of calcium is maintained by a voltage-gated calcium channel, whereas intracellular calcium is maintained by the calcium stored in the cell like the endoplasmic reticulum [7]. The physiological studies show that there are impairments in the normal functioning of the calcium-binding buffers, voltage-gated channel, and endoplasmic reticulum when the cell suffers from neurodegenerativity [7]. The level of buffers is found to decrease in certain parts of the brain i.e., hippocampal and basal forebrain regions, [2], [16], [17] whereas there is a noteworthy increase in the voltage-dependent calcium channels and fluxes through the endoplasmic reticulum, which results in the higher amount of cytosolic calcium in the cell [18]–[20]. This higher level of calcium in the cell renders toxicity to the cell which makes the neuron vulnerable to Alzheimer's disease. The sudden decrease and increase of buffer concentrations, voltage-gated calcium channels, and endoplasmic reticulum fluxes respectively result in the alteration of calcium homeostasis which directly affects the cell fate. It has been found that various buffers like calmodulin and calbindin-D_{28k} have their hands in the progression of this dementia, whereas, the type of voltage-gated calcium channel may have its active participation in Alzheimer's is the L-type of calcium channel whose functions are altered in it.

3.2 Literature Survey

In the last few years, the research edifying the computational modeling of the physiological phenomenon of calcium concentration distribution has emerged on a large scale. In the years 2013 and 2014, Jha et al. have proposed the said model in presence of VGCC as well as buffer and buffer respectively [21],[22] . They have used two-dimensional finite element model for astrocytes to obtain the desired results. In 2014, Kotwani and Adlakha have used a two -dimensional computational models to show the flow of calcium in fibroblast cells having calcium-binding protein [23]. Also, the duo of Jha and Adlakha has been working on mathematical modeling of calcium concentration distribution in neuronal cells. They have delineated calcium flow in presence of buffers and used the finite element method for the two-dimensional mathematical models in cartesian co-ordinates [24]. Jha and Adlakha have extended their model in 2015, with polar coordinates, and studied the combined effect of ER leak, SERCA, and EGTA on calcium concentration distribution using the finite element method [25]. They have used a circular domain as approximate geometry to obtain the results. Also, the other researchers have proposed one-dimensional and two-dimensional mathematical models portraying the calcium diffusion in presence of buffers and ER for various cells like neurons, oocytes, and cardiac myocytes [26]–[30]. Similarly, Naik and Pardasani have studied the mathematical physiology of calcium flow in cardiac myocytes. For the same, they have used a one-dimensional finite element model in presence of ER and buffer [31]. Further, they have extended their study in two dimensions in presence of buffers and solved using the finite element method [32].

Thus, on the basis of the literature survey, it is found that none of the authors have considered neurodegenerative disorders and the irregular geometry of the cell. Thus, an attempt has been in this research article to study the impact of calcium concentration distribution in presence of buffers, voltage-gated calcium channels, and endoplasmic reticulum in normal and Alzheimer's affected neuronal cells. For this, a two-dimensional mathematical model has been constructed by incorporating the parameters like diffusion co-efficient, association rate of buffers, voltage-gated channel, and endoplasmic reticulum-related leak and pumps. Further, this computational model is treated with the finite element technique and the results are simulated using MATLAB.

3.3 Mathematical Formulations

Mathematical modeling of the physiological phenomena taking place in the neuronal cell of the most complex live structure of the world i.e. brain is done using a partial differential equation. The two-dimensional computational models delineating the calcium diffusion are constructed in this section. But before it, the mathematical formulation of the parameters considered here, i.e., buffers, voltage-gated calcium channels, and endoplasmic reticulum is shown. Further, the model is proposed followed by the boundary conditions matching the physiological conditions of the brain.

3.3.1 Calcium Buffering

It is known that almost 99% of the calcium gets bound at the cytosolic level and only the rest amount of the calcium is used for further functioning [15]. Hence, the role of buffers in maintaining the cell is immense. Physiologically, calcium after entering the cytosol reacts with buffers, and as result calcium bound buffers are formed. Mathematically it can be stated as [33]–[35]:

$$[Ca^{+2}] + [B_j] = [CaB_J] \tag{3.1}$$

where Ca^{+2} is the cytosolic calcium concentration, $[B_j]$ is the buffer concentration and $[CaB_J]$ is the concentration of the calcium bound buffers present at the cytosolic level for buffer j.

Using law of mass action and Fick's law, the diff'rential equation stating the change in calcium, buffer and calcium bound buffers are shown as [36]

$$\frac{\partial C}{\partial t} = D_{Ca} \left(\frac{\partial^2 C}{\partial x^2} + \frac{\partial^2 C}{\partial y^2} \right) + \sum_j R_j + \sigma_{Ca} \tag{3.2}$$

$$\frac{\partial [B_j]}{\partial t} = D_{B_j} \left(\frac{\partial^2 [B_j]}{\partial x^2} + \frac{\partial^2 [B_j]}{\partial y^2} \right) + \sum_j R_j \tag{3.3}$$

$$\frac{\partial [CaB_j]}{\partial t} = D_{CaB_j} \left(\frac{\partial^2 [CaB_j]}{\partial x^2} + \frac{\partial^2 [B_j]}{\partial y^2} \right) - \sum_j R_j \tag{3.4}$$

where D_{Ca}, D_{B_j} and D_{CaB_j} are diffusion coefficients of calcium, free buffer and calcium bound buffer respectively and

$$R_j = -k_j^+ [B_j] [Ca^{+2}] + k_j^- [CaB_j] \tag{3.5}$$

where R_j is the reaction term for buffer j.

3.3.2 Voltage Gated Calcium Channel (VGCC)

The calcium outside the cells has several ways of entering the domain of the cell. Out of all, VGCC is the principal route for calcium influx in neuronal cells. There are totally five types of voltage-dependent/ gated calcium channels, i.e., P, Q, N, T, and L [37]. Each of these channels has a specific location and role to perform. It has been found that out of all, L-VGCC's are responsible for pertrbation in normal neuronal calcium homeostasis. They are located in the bodies of the neuron cells; dendrites and spines. It has 4 sub-types, CaV 1.1 1.4. L- VGCC's have been found to regulte neuronal activities which further leads to behavioral changes [37]. Also, It has been found that there is a striking increase in activities of L- type calcium channels in a broad spectrum of CNS disorders [37], [38].

The formulation of voltage-gated calcium channel is done mathematically using Goldman Hodgkin Katz (GHK) current which can be formulated as [21], [35]

$$I_{ca} = P_{ca}z_{ca}^2 \frac{F^2 V_m}{RT} \frac{[Ca^{+2}]_i - [Ca_0^{+2} \exp\exp(-z_{ca}\frac{FV_m}{RT})}{1 - \exp\exp(-z_{ca}\frac{FV_m}{RT})} \qquad (3.6)$$

Further, the flux of the calcium at the cytosolic level is regulated by converting the above-mentioned equation into the form of σ. The GHK current equation can be converted into moles/ second equation as [21], [39]

$$\sigma_{Ca} = \frac{-I_{Ca}}{z_{Ca}FV_{nervecells}} \qquad (3.7)$$

3.3.3 Endoplasmic Reticulum (ER)

The endoplasmic reticulum is present in almost all eukaryotic cells. Generally, it consists of half of the cell volume and is known as the storehouse of calcium as it contains the highest amount of calcium. There are basically two types of endoplasmic reticulum, rough and smooth. Smooth endoplasmic reticulum and its modified version, the sarcoplasmic reticulum is responsible for the efflux and in-flux of calcium in ER from cytosol [40]. During the mathematical formulation of the endoplasmic reticulum, the calcium buffering of the endoplasmic reticulum calcium is neglected over here. For formulating the calcium dynamics equation, the model of De-Young Keizer is used and has been followed [41]. The conservation of calcium can be stated as, [42]

$$C_1 = \frac{V_{ER}}{V_{Cyt}}, \ C_0 = C_1[Ca^{+2}]_{ER} + [Ca^{+2}]_i, \ [Ca^{+2}]_{ER}$$

$$= \frac{\eta_{ER}}{V_{ER}}, \ [Ca^{+2}]_i = \frac{\eta_{Cyt}}{V_{Cyt}} \qquad (3.8)$$

where C0 is the total calcium concentration, C1 is the volume ratio. Further, the leaks and pumps which are taken into consideration for calcium flow between endoplasmic reticulum and cytosol are J_{leak}, J_{chan} and J_{pump}. The material balance equation is stated as [41]

$$\frac{d[Ca^+]_i}{dt} = C_1 [J_{leak} + J_{chan}] - J_{pump} \qquad (3.9)$$

Further, the formulation for pump, leak and channel portraying the calcium diffusion between endoplasmic reticulum and cytosol can be

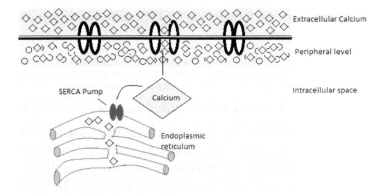

Figure 3.1 Calcium diffusion in presence of buffers

done as [35], [42]:

$$J_{pump} = P_R^{max} \frac{[Ca^{2+}]_i^2}{[Ca^{2+}]_i^2 + (K_R^M)^2} \tag{3.10}$$

$$J_{leak} = \frac{D_{leak}}{C_1}(1 + C_1)(\frac{C_0}{1 + C_1} - [Ca^{2+}]_i) \tag{3.11}$$

$$J_{chan} = \frac{D_{chan}}{C_1}(1 + C_1)(\frac{C_0}{1 + C_1} - [Ca^{2+}]_i) \tag{3.12}$$

Where P_R^{max} is the maximum pump rate and K_R^M is the Michaelis-Menten constant, D_{leak} is the diffusivity constant and D_{chan} is the conductance of the channel. The activities between ER and cytosol is increased significantly in

The physiology of calcium diffusion in presence of buffers, VGCC and ER can be visualized as shown in Figure 3.1

VGCC and ER. Calcium from the extra- cellular space enters into cytosol via VGCC where it gets bound with the calcium binding buffers. These buffers are present at the peripheral level. Further, calcium from ER is managed by SERCA pump and thus the cell calcium is maintained.

Considering the equations, (3.2), (3.5),(3.7), and (3.9-3.12), the proposed two-dimensional mathematical model is stated as

$$\frac{\partial C}{\partial t} = D_{Ca}\frac{\partial^2 C}{\partial x^2} + D_{Ca}\frac{\partial^2 C}{\partial y^2}(1 + C_1)\left(D_{leak} + PD_{chan}\right)\left(\frac{C_0}{1 + C_1} - C\right)$$

$$- P_R^{max}\left(\frac{C^2}{C^2 + K^2}\right) - K^+ B_\infty[C - C_\infty] \tag{3.13}$$

In the above equation, the formulation for ER can be can further divided into two cases, [42],[43] i.e.

$$Case\ 1: For\ K \gg C, \frac{C^2}{C^2+K^2} < \frac{C^2}{K^2} < \frac{C}{K} \qquad (3.14)$$

$$Case\ 2: For\ K \ll C, K = \alpha C; 0 < \alpha < 1 \qquad (3.15)$$

For further employment of the numerical technique and to solve the above partial differential equation of calcium concentration distribution, the boundary conditions matching the physiology are taken as follows. The boundary conditions are considered here at different points of the boundary of the neuron cells in the form of Neumann boundary conditions

$$[Ca^{+2}]_{t=0} = 0.1 \mu M \qquad (3.16)$$

$$[Ca^{+2}]_\infty = 0.1 \mu M \qquad (3.17)$$

$$-D_{Ca} \frac{\partial C}{\partial \eta} = \sigma_{Ca} + \sigma_{VGCC} \qquad (3.18)$$

Thus, having formulated the mathematical model and being equipped with the necessary boundary condition, the finite element method is employed to obtain the desired results, showing calcium diffusion in normal and Alzheimeric neurons.

3.4 The Finite Element Technique

The use of the finite element method began in various fields of science and technology in 1960 [44]. The method yields approximate values of the unknowns at a discrete number of points over the domain. It gives a better approximate solution to the irregular and complex geometries [45] Over and above this, different properties can be incorporated into a single structure. A retrospective survey shows that the finite element method has a wide range of applications in almost all fields of biology. Agrawal et al. have used the finite element method to investigate the thermal changes in human limbs[46]. They have studied the impact and relation of various physiological parameters in two-dimensional cases to get a better view of the thermal alterations. Makrariya and Adlakha have been working on the mathematical modeling of the temperature distribution in dermal tissues of spherical-shaped organs [47]. They have extended their study of a thermal investigation by studying the thermal conductivity in human breast having uniformly perfused tumors [48]. The finite element method has been adopted by them to study the above-stated two-dimensional steady-state computational models. Whereas, Khanday and Saxena have used the finite element method for the mathematically modeled physiological phenomenon of heat regulation in the human head [49]. They had proposed one dimensional model for a transient state which was based on a bio-heat equation and the changes are analyzed at micro levels for the precise outcome. Further, Khanday et al. extended their visualization of fluid distribution in human dermal regions on the basis of the radial diffusion equation [50]. The finite element method had been employed by them to obtain the results for fluid concentration distribution at dermal and sub-dermal regions. Acharya et al. also contributed to the study of temperature distribution in male and female skin layers [51]. They have used the finite element method for solving the mathematical model involving bio-heat equations. Thus, we can conclude that FEM has been extensively used for portraying various physiological phenomena of biology. Over and above this, analytically, [52]–[54] as well as numerically, the finite element is used widely to study the impact of calcium signaling in various cells in presence of certain parameters. But those researchers have employed approaches like variational Rayleigh-Ritz or Galerkin approximation and that too solely considering a regular-shaped domain. The irregular geometry of the cell and the physiological fact of the flux through multiple boundaries have been ignored and are done rarely by researchers [55], [56] . Thus, based on it and various computational and simulation-based research, [57], [58], [59], in this chapter, we have considered the irregular geometry of a typical neuron cell having multiple boundaries through which the calcium influx can take place. The geometries of the neuron cells analyzed in the present chapter cannot be handled using analytical as well as finite difference methods. Moreover, for the betterment of the approximate solution obtained, refinement of the initial elements is done. The quality of the mesh generated is also checked and shown in the following subsection.

3.4.1 Approximated Geometry of the Cell

The targeted domain is the approximate geometry of the neurons. The shape and size of the neurons vary according to the region of the brain in which it is located. Here, we have considered the hippocampal and the basal forebrain regions as these regions are damaged highly in dementia named Alzheimer's disease [2], [60], the approximate geometry of hippocampal neurons is shown in Figure 3.2. The hippocampal neuron is pyramidal in nature having a

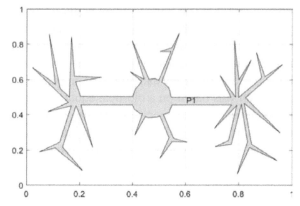

Figure 3.2   Approximated structure of a hippocampal neuron

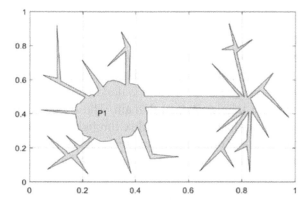

Figure 3.3   Approximated structure of a basal forebrain neuron

number of dendrites and axon terminals [61]. Here, we have considered the neuron having a number of dendrites, soma - the cell body, and axon terminals.

The geometry of the basal forebrain neuron is shown in Figure 3.3. Unlike hippocampal neurons, it is multipolar in nature. It consists of several dendrites, soma, and axon terminals.

3.4.2 Physiological Boundary Conditions

The Neumann boundary condition which delineates the calcium flux is shown structurally. Figures 3.4 and 3.5 show the calcium entry/ entries in the cytosol of both, hippocampal and basal forebrain neuronal cells through the sole and

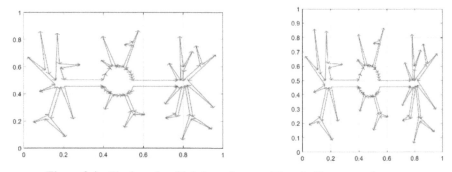

Figure 3.4 Single and multiple boundary conditions in hippocampal neuron

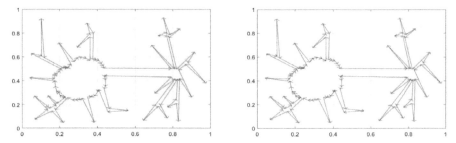

Figure 3.5 Single and multiple boundary conditions in basal forebrain neuron.

multiple channels. The reason for taking multiple fluxes is the physiological background in the brain. Physiologically, the calcium is entered at the periphery region were it gets sequestered. Multiple channels have been considered and shown by Jha et al. by considering the line source of calcium ions at the boundary of the cytosol in x and y directions [15]. In Figures 3.4 and 3.5, multiple channels are incorporated directly in place of considering the line source.

3.4.3 Meshing of the domain

Figures 3.6 and 3.7 shows the initial and the refined meshing of the hippocampal and the basal forebrain neurons respectively. The refinement is done to check its sensitivity towards the mesh. Here the domain of the hippocampal neuron is initially discretized into 1374 triangles having 1112 nodes, whereas, on refinement, the number of triangles is increased to 5496 triangles and 3597 nodes. In the same manner, the basal forebrain neuron is divided into 1486 triangles initially having 1111 nodes. Further, refining the mesh leads to 5944

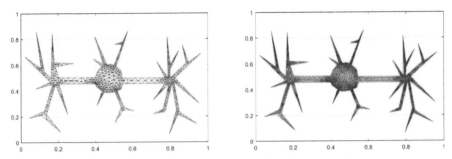

Figure 3.6 Initial and refined meshing of the hippocampal neuron

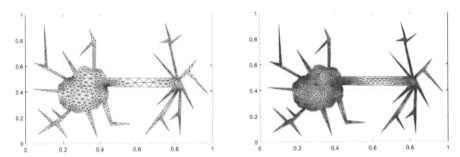

Figure 3.7 Initial and refined meshing of the basal forebrain neuron

triangles and 3707 nodes. Purpose of refining the mesh in irregular geometry is to increase the triangle quality as shown in Figures 3.8 and 3.9

It has been suggested by Bank that if the value of q is above 0.6, the triangle quality is acceptable, [62] where the value of q is calculated using the below formula

$$q = \frac{4a\sqrt{3}}{h_1^2 + h_2^2 + h_3^2} \tag{3.19}$$

Where a is the area h_1, h_2 and h_3 are the side lengths of the triangle. From both the above figures it is observed that the mesh quality is highly acceptable for refined cases in hippocampal as well as basal forebrain neurons.

It has been observed from the refinement of the mesh and the triangle qualities of both the meshes that the refined mesh shows more appropriate mesh generation and hence a better approximate result can be obtained for the same

Hence, in the next section, the results are obtained only for the refined mesh and not the initial meshing.

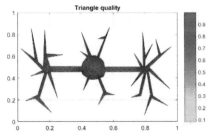
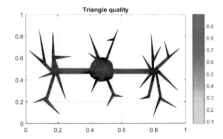

Figure 3.8 Triangle quality for initial and refined mesh in hippocampal neuron

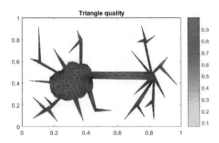
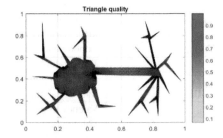

Figure 3.9 Triangle quality for initial and refined mesh in basal forebrain neuron

3.5 Results and Discussion

While considering the physiological parameters, the varying computational values used to obtain the results are shown in Table 3.1 or else stated when and where necessary.

3.5.1 For Hippocampal Neuron

Figure 3.10 depicts the calcium diffusion in normal and Alzheimeric cells having refined mesh. The single flux of calcium ions is considered. By considering both the cases of endoplasmic reticulum, the results obtained are moreover the same, i.e., there is no significant difference found. The results are obtained by considering the normal cell, i.e., in presence of a sufficient amount of calmodulin buffer present in the hippocampal area. Further, the results are obtained by considering the Alzheimeric cell, [2] i.e., in presence of 20-30% of total calmodulin buffer. It is observed from the above graphs that due to an increase in the voltage-gated calcium flux and inward flux from the endoplasmic reticulum, there is an increase in calcium concentration. Both the cases of endoplasmic reticulum are considered over here. Comparing both the figures, it is found that the amount of calcium ions is increased which leads to neurodegenerative of the cell and hence Alzheimer's disease.

Table 3.1 Value of physiological parameters [29],[32], [39],[42],[52],[63]

Symbol	Parameter	Value
D	Diffusion coefficient	200-300 $\mu m^2 s^{-1}$
k^+Calmodulin	Buffer association rate	120 $\mu m^{-1} s^{-1}$
k^+ Calbindin	Buffer association rate	75 $\mu m^{-1} s^{-1}$
$[B]_\infty$ Calmodulin	Buffer concentration	2-25 μM
$[B]_\infty$Calbindin	Buffer concentration	100-350 μM
$[Ca^{2+}_\infty]$	Background calcium concentration	0.1μM
V	Volume of cytosol	5.233*10^13 l
D_{leak}	Diffusivity constant	0.11
F	Faraday's constant	96,485 C/mol
C_0	Total calcium concentration	1
R	Ideal gas constant	8.31 J/(mol*k)
D_{chan}	Conductance of the channel	6
T	Temperature	300k
$P_R{}^{max}$	Maximum pump rate	0.9
P_{oout}	Rate of calcium efflux from cytosol	$0.5s^{-1}$
Z_{Ca}	Valency of calcium ion	2
C_1	Volume ratio	0.185 μM
K	Michaelis-Menten constant	0.1

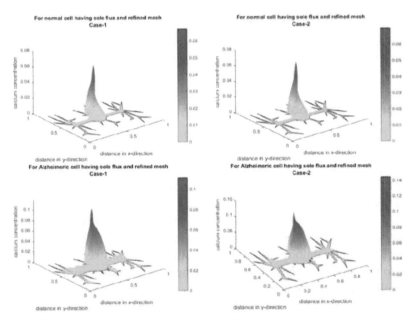

Figure 3.10 Calcium concentration distribution for case 1 and case 2 in normal and Alzheimeric cells having single flux

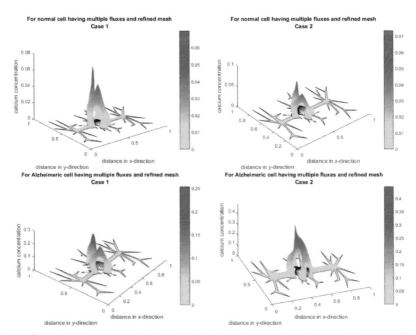

Figure 3.11 Calcium concentration distribution for case 1 and case 2 in normal and Alzheimeric cells having multiple fluxes

Figure 3.11 shows the estimated calcium distribution in normal and diseased cells having multiple boundaries. Here both the cases of endoplasmic reticulum are used to obtain the results. The reason for taking the multiple boundaries is the physiological view of the study. Due to the multiple fluxes, more area of the cell is affected by the higher level of calcium and also there is an increase in the level of the calcium in comparison to the single flux. It is observed that due to the fluxes at the boundary the calcium in that area is maximum and finally attains the background concentration. In these graphs also, the significant change in the calcium in Alzheimer's affected cells can be seen which may further lead to the loss of neurons and lack of memory formation in that particular area of the brain.

3.5.2 For Basal Forebrain Neuron

Figure 3.12 shows the calcium concentration distribution in basal forebrain neurons having a single calcium flux. As stated in the above figure, both cases are also taken into consideration. In the basal forebrain neuron, the buffer

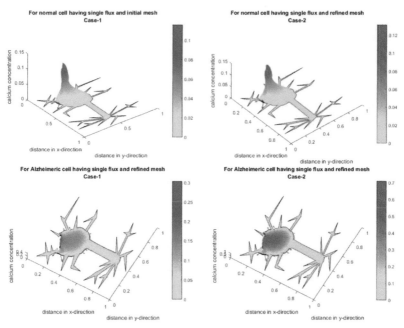

Figure 3.12 Calcium concentration distribution for case 1 and case 2 in normal and Alzheimeric cells having single flux

which plays a crucial role is calbindin. Due to a decrease in the buffer amount and increase in the flux of calcium via endoplasmic reticulum and voltage-gated calcium channels in basal forebrain neurons, a significant hike in the level of calcium concentration is observed. Physiologically, the neurons of the basal forebrain are the first to get affected by Alzheimeric changes which take place with pre-tangle formation due to an increase in calcium [64]. Here a significant change in normal and Alzheimeric cases is seen.

Figure 3.13 shows the calcium concentration distribution in basal forebrain neurons in presence of calbindin having multiple calcium fluxes for case-1 and case-2. Due to multiple fluxes, there is a change in the somatic region, i.e., more area is affected which can be seen in comparison with Figure 3.12.

Over and above the individual results of hippocampal and basal forebrain neurons, if we compare the results for both the neurons, then it can be found on the observation that the fact of the loss of basal forebrain neurons primarily is verified. It can be seen that the calcium concentration level is higher in basal forebrain neurons which makes them vulnerable to tangle formation and hence the neurons are lost first in this area [16], [60].

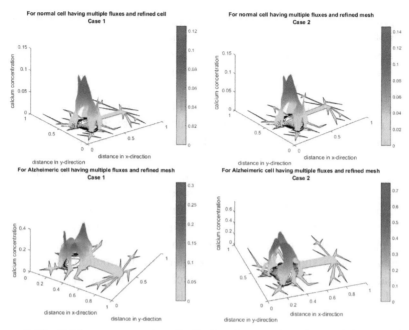

Figure 3.13 Calcium concentration distribution for case 1 and case 2 in normal and Alzheimeric cells having multiple fluxes.

3.6 Conclusion

In this chapter, an attempt has been made by us to delineate the physiological process of calcium concentration distribution taking place in normal and Alzheimeric neurons- which are considered the building blocks of the central nervous system. Till date, no efforts have been made to computationally delineate the physiological process in irregularly shaped geometry of the cell.

For this, a two-dimensional mathematical model has been developed in presence of parameters like the calcium-binding proteins, voltage-gated calcium channel, and endoplasmic reticulum. The conditions matching the physiology of the brain cells have been assumed in presence of other important parameters like diffusion coefficient of the calcium, voltage-gated calcium flux and leak, pump, and channel associated with the endoplasmic reticulum. To solve the mathematical model and obtain the results, the finite element technique has been employed over here. The obtained results clearly show the significant impact of buffers, ER, and VGCC on the calcium concentration. Also, it has been found that the basic nature of the calcium flow in normal conditions of the cell is the same when compared with some previously solved finite element results [21], [25]. A significant change has been observed in the flow of calcium in normal and Alzheimeric cells. The basic idea is the hike in the calcium concentration in Alzheimeric cells has been verified over here. The higher level of calcium leads to the toxicity of the cell and the cell becomes vulnerable to Alzheimer's which further results in the formation of the tangles as well as the pre-tangles in the basal forebrain neuron. Also, the increase in the calcium leads to neuronal loss in the hippocampal area which further leads to the alterations in the memory formation in that particular area. Also, to obtain better approximate results, the refinement of the mesh has been done. On the basis of this mathematical model, the estimated flow of the calcium can be predicted in the Alzheimeric condition of the neurons which can further help the theoretical scientists and researchers working in this field. Further, this kind of mathematical model can be extended in the form of three dimensions and in presence of more diverse parameters having a significant impact on calcium distribution which may further lead to a more precise view of the calcium concentration distribution in normal as well as Alzheimer's affected cells.

References

[1] Laferla FM, Calcium dyshomeostasis and intracellular signalling in Alzheimer's disease, Nat'Rev Neurosci 3: 862-872, 2002.

[2] Turkington C, Mitchell D, The Encyclopedia of Alzheimer's Disease, Sec'nd. Facts On File: An imprint of Infobase Publishing, 2010.

[3] Magi S et al., Intracellular calcium dysregulation: Implications for Alzheimer's disease, Bio'ed Res. Int. 1-14, 2016.

[4] Abramov AY, Canevari L, Duchen MR, Calcium signals induced by amyloid β peptide and their consequences in neurons and astrocytes in culture, Biochim. Biophys. Acta 1742: 81-87, 2004.

[5] Rapaka D et al., Calcium regulation and Alzheimer's disease, Asian Pacific J. Trop. Dis.4: S513-S518, 2014.

[6] Brzyska M, Elbaum D, Dysregulation of calcium in Alzheimer's disease, Acta Neurobiol. Exp. 63: 171-183, 2003.

[7] Brawek B, Garaschuk O, Network-wide dysregulation of calcium homeostasis in Alzheimer's disease, Cell Tissue Res 357: 1–12, 2014.

[8] Mattson MP, Chan SL, Neuronal and glial calcium signaling in Alzheimer's disease, Cell Calcium 34: 385-397, 2003.

[9] Kawamoto EM , Vivar C, Camandola S, Physiology and pathology of calcium signaling in the brain, Front. Pharmacol. 3: 1-17, 2012.

[10] Nikoletopoulou V, Tavernarakis N, Calcium homeostasis in aging neurons, Front. Genet. 3: 1-17, 2012.

[11] Korol TY et al., Disruption of calcium homeostasis in Alzheimer's disease, Neurophys- iology 40: 457-464, 2008.

[12] Khachaturian Z, Introduction and overview, Ann. New York Acad. Sci. 4-7, 1989.

[13] Green KN, Laferla FM, Linking calcium to A β and Alzheimer's disease, Neuron 59: 190-194, 2008.

[14] Khachaturian Z, Calcium hypothesis of Alzheimer's disease and brain aging, Ann. New York Acad. Sci. 1-11, 1993.

[15] Jha,A, Adlakha N, Jha BK, Finite element model to study effect of Na+-Ca2+ ex- changers and source geometry on calcium dynamics in a neuron cell, J. Mech. Med. Biol. 16: 1-22, 2015.

[16] Geula C et al., Loss of Calbindin-D_{28K} from aging human cholinergic basal forebrain: Relation to plaques and tangles, J. Neuropathol. Exp. Neurol. 6262: 605-616, 2003.

[17] Ahmadian S et al., Loss of Calbindin-D_{28K} is associated with the full range of tangle pathology within basal forebrain cholinergic neurons in Alzheimer's disease, Neurobiol. Aging 36: 3163–3170, 2015.

[18] Li J et al., Endoplasmic reticulum dysfunction in Alzheimer's disease, Mol Neurobiol 51: 383–395, 2014.

[19] Popugaeva E et al., Role of endoplasmic reticulum Ca2+ signaling in the pathogenesis of Alzheimer's disease, Front. Mol. Neurosci. 6: 1-7, 2013.

[20] Anekonda TS et al., Neurobiology of disease L-type voltage-gated calcium channel block- ade with isradipine as a therapeutic strategy for Alzheimer's disease, Neurobiol. Dis. 41: 62-70, 2011.

[21] Jha BK, Adlakha N, Mehta MN, Two-dimensional finite element model to study cal- cium distribution in astrocytes in presence of VGCC and excess buffer, Int. J. Model. Simulation, Sci. Comput. 4: 12500301-125003015, 2013.

[22] Jha BK, Adlakha N, Mehta MN, Two-dimensional finite element model to study calcium distribution in astrocytes in presence of excess buffer, Int. J. Biomath. 7: 1-11, 2014.

[23] Kotwani M, Adlakha N, Mehta MN, Finite element model to study the effect of buffers, source amplitude and source geometry on spatiotemporal calcium distribution in fibroblast cell, J. Med. Imaging Heal. Informatics 4: 840-847, 2014.

[24] Jha A, Adlakha N, Finite element model to study the effect of exogenous buffer on calcium dynamics in dendritic spines, Int. J. Model. Simulation, Sci. Comput. 5: 1-12, 2014.

[25] Jha A, Adlakha N, Two-dimensional finite element model to study unsteady state Ca2+ diffusion in neuron involving ER, LEAK and SERCA, Int. J. Biomath. 8: 1-14, 2015.

[26] Tewari SG, Pardasani KR, Finite element model to study two dimensional unsteady state cytosolic calcium diffusion in presence of excess buffers, IAENG Int. J. Appl. Math. 40: 1-5, 2010.

[27] Tewari SG, Pardasani KR, Finite element model to study two dimensional unsteady state cytosolic calcium diffusion, J. Appl. Math. Informatics 29: 427-442, 2011.

[28] Panday S, Pardasani KR, Finite element model to study effect of buffers along with leak from ER on cytosolic Ca2+ distribution in oocyte, IOSR J. Math. 4: 1-8, 2013.

[29] Pathak K, Adlakha N, Finite element model to study calcium signalling in cardiac myocytes involving pump, leak and excess buffer, J. Med. Imaging Heal. Informatics 5: 1-6, 2015.

[30] Pathak K, Adlakha N, Finite element model to study two dimensional unsteady state calcium distribution in cardiac myocytes, Alexandria J. Med. 52: 261-268, 2016.

[31] Naik PA, Pardasani KR, One dimensional finite element model to study calcium distri- bution in oocytes in presence of VGCC, RyR and buffers, J. Med. Imaging Heal. Infor- matics 5: 471-476, 2015.

[32] Naik PA, Pardasani KR, Finite element model to study calcium distribution in oocytes involving voltage gated Ca2+ channel, ryanodine receptor and buffers, Alexandria J. Med. 52: 43-49, 2016.

[33] Smith GD, Analytical steady-state solution to the rapid buffering approximation near an open Ca2+ channel, Biophys. J. 71: 3064-3072, 1996.

[34] Smith GD et al., Asymptotic analysis of buffered calcium diffusion near a point source, Siam J. Appl. Math. 61: 1816-1838, 2001.

[35] Keener J, Sneyd J, Mathematical Physiology, Second. Springer US, 2009.

[36] Crank J, The Mathematics of Diffusion, Second. Clarendon Press Oxford, 1975.

[37] Yagami T, Kohma H, Yamamoto Y, L-type voltage-dependent calcium channels as therapeutic targets for neuro- degenerative diseases, Curr. Med. Chem. 1: 4816-4827, 2012.

[38] Schampel A, Kuerten S, Danger: High voltage The role of voltage-gated system pathol- ogy, Cells 6: 1-8, 2017.

[39] Dave DD, Jha BK, Modeling the alterations in calcium homeostasis in the presence of protein and VGCC for Alzheimeric cell, in M. Pant, K. Ray, T. Sharma, S. Rawat, and A. Bandyopadhyay (Eds.), Advances in Intelligent Systems and Computing, Springer, pp. 181-189, 2018.

[40] Rajagopal S, Ponnusamy M, Calcium signaling: From physiology to diseases Springer Nature Singapore Pvt Ltd., 2017.

[41] Youngt G, Keizer J, A single-pool inositol 1,4,5-trisphosphate-receptor-based model for agonist-stimulated oscillations in Ca2+ concentration, Proc. Natl. Acad. Sci. USA 89: 9895-9899, 1992.

[42] Jha BK, Adlakha N, Mehta MN, Finite volume model to study the effect of ER flux on cytosolic calcium distribution in astrocytes, J. Comput. 3: 74-80, 2011.

[43] Tripathi A, Adlakha N, Finite volume model to study calcium diffusion in neuron in- volving J_{RYR}, J_{SERCA} and J_{leak}, J. Comput. 3: 41-47, 2011.

[44] Rao SS, The Finite Element Method in Engineering, 5th ed. Elsevier, 2011.

[45] Seshu P, Textbook of Finite Element Analysis, PHI Learning Private Limited, 2012.

[46] Agrawal M, Adlakha N, Finite element model to study thermal effect of uniformly perfused tumor in dermal layers of elliptical shaped human limb, Int. J. Biomath. 4: 241-254, 2011.

[47] Khanday MA, Saxena VP, Finite element estimation of one-dimensional unsteady state heat regulation in human head exposed to cold environ- ment, J. Biol. Syst. 17: 853-863, 2009.

[48] Khanday MA, Aijaz M, Rafiq A, Numerical estimation of the fluid dis- tribution pattern in human dermal regions with heterogeneous metabolic fluid generation, J. Mech. Med. Biol. 15: 15500011-155000112, 2015.

[49] Acharya S, Gurung DB, Saxena VP, Two-dimensional finite element method for metabolic effect in thermoregulation on human males and female skin layers, J. Coast. Life Med. 3: 623-629, 2015.

[50] Dave DD, Jha BK, Delineation of calcium diffusion in alzheimeric brain, J. Mech. Med. Biol. 18: 1-15, 2018.

[51] Jha A, Adlakha N, Analytical solution of two-dimensional unsteady state problem of calcium diffusion in a neuron cell, J. Med. Imaging Heal. Informatics 4: 547-553, 2014.

[52] Jha BK, Adlakha N, Mehta MN, Analytic solution of two-dimensional advection diffusion equation arising in cytosolic calcium concentration distribution, Int. Math. Forum 7: 135-144, 2012.

[53] Khalid et al., Geometry-based computational modeling of calcium signaling in an astrocyte, in IFMBE Proceedings 65: 1-4, 2018.

[54] Jha BK, Jha A, Adlakha N, Finite element estimation of calcium ions in presence of NCX and buffer in astrocytes, Int. J. Pharma Med. Biol. Sci. 5: 1-5, 2016.

[55] Ruppin E, Reggia JA, Cortical spreading depression and the pathogenesis of brain disorders: A computational and neural network-based investigation, Neurol. Res. 2: 447- 456, 2001.

[56] Saxena P, Goldstein R, Isaac L, Computer simulation of neuronal toxicity in the spinal cord, Neurol. Res. 19: 340-348, 2016.

[57] Jha BK and Dave DD, Approximation of Calcium Diffusion in Alzheimeric cell, Journal of Multiscale Modelling, Accepeted Manuscript.

[58] Riascos D et al., Age-related loss of calcium buffering and selective neuronal vulnera- bility in Alzheimer's disease, Acta Neuropathol 122: 565-576, 2011.

[59] Lodish H et al., Molecular Cell Biology, Seventh. New York: W. H. Freeman and Company, 2013.

[60] Bank RE, A Software Package for Solving Elliptic Partial Differential Equations Users Guide 12.0, 2016.

[61] Bertram R et al., A simplified model for mitochondrial ATP production, J. Theor. Biol. 243: 575-586, 2006.

[62] Duan L et al., Stem cell derived basal forebrain cholinergic neurons from Alzheimer's disease patients are more susceptible to cell death, Mol. Deger. 1-14, 2014.

4

Comparative Analysis of Computational Methods used in Protein–Protein Interaction (PPI) Studies

Khushi Ahuja, Aditi Joshi, Navjyoti Chakraborty, Ram Singh Purty, and Sayan Chatterjee

University School of Biotechnology, Guru Gobind Singh Indraprastha University, New Delhi, India
E-mail: sayan@ipu.ac.in

Abstract

Proteins are the functional molecules of cells involved in catering to most of the biological processes. Different proteins interact with each other in various combinations to regulate and execute these functions, thus identification of Protein–protein interactions (PPIs) is crucial for understanding regulatory and metabolic pathways. *In vivo* and *in vitro* experiments like pull-down, tandem affinity purification, co-immunoprecipitation, Yeast 2-hybrid, etc. have been used traditionally to identify and validate PPIs but are cumbersome and time-consuming. Given these limitations, *in silico* or computational approaches provide an easy and time-saving approach to identify the PPIs, which use an array of biological data like 3D structure-based data, sequence data, expression patterns, and evolution data. The statistical rigor of the analysis is also used along with computational methods for the identification of PPIs to further consolidate the data obtained. In this chapter, we have listed, reviewed, and compared different available methods, servers, databases, and tools used for the prediction and identification of PPIs. Being the building blocks of all living organisms, proteins and protein interaction become an important aspect of the working of entire nature, from the functioning of an organ system to the functioning of the organelles at the lowest cellular level. The protein interactions constitute a great deal of what is still unknown. An area where bioinformatics has paved a way is to predict interactions which may then even be validated further *via* wet-lab experiments. The paper focuses on compiling the current bioinformatics methods that help in predicting and analyzing protein interactions at the molecular level among different organisms.

Keywords: proteins, computational methods, PPI, biological databases, molecular docking.

4.1 Introduction

4.1.1 Protein

Proteins are biological macromolecules that are made of one or more polypeptide chains. Each polypeptide is built up of a chain of amino acids linked together by peptide (chemically amide) bonds while the exact amino acid sequence is determined using the genes coding for that specific polypeptide. Amino acids have Hydrogen (H), Oxygen (O), and Nitrogen (N) atoms combined with different functional groups, all amino acids having the same functional groups except a variable R group. To overcome the hydrophobic collapse, the protein folds into secondary and tertiary structures. Protein folding, as well as Protein–protein interactions, are based on the non-covalent interactions (like weak van der Waal forces, hydrophobic interactions, hydrogen bonding) between the residue side chains. In diverse pathways involved in many biological functions, proteins interact with other small proteins.

4.1.2 Protein–Protein Interaction

It is very rare for proteins to operate alone as their functions are often regulated. Various molecular processes inside a cell performed by molecular machines are made using several protein components organized using their Protein–Protein Interactions (PPIs) and are called Protein–protein complexes. Such PPIs form the organism's Interactomics while abnormal PPIs are the usual basis of diverse aggregation-related diseases, for instance, Alzheimer's disease, Creutzfeldt Jakob disease, etc.

PPIs are termed as the physical unions of high specificity formed between a minimum of two or more two protein molecules. This is a result of biochemical events influenced by interactions like hydrogen bonding, hydrophobic effect, and electrostatic forces. Examples of PPIs in the bodies of organisms include electron transfer proteins, signal transduction proteins, proteins in cell metabolism, and proteins in muscle contraction to name a few.

Homocomplexes and heterocomplexes are two types of Protein–protein complexes. Homocomplexes, such as cytochrome c', is usually optimized and permanent in nature, whereas heterocomplexes, such as antibody complex HYHEL-5 with lysozyme, can be formed and destroyed depending on the surroundings or other exterior components and include proteins that can manage independently as well.

4.1.2.1 Protein–protein interfacial characteristics

Many fundamental aspects are used to characterize a Protein–protein interface. These are usually calculated through the coordinates of the complex. These include characteristics like size, shape, and hydrophobicity that have been discussed in the upcoming sections.

4.1.2.1.1 Size and shape

The size and shape of protein interfaces are often expressed in Angstroms (Å) or, more precisely, in solvent accessible surface area (ASA), which is based on the known correlation between the hydrophobic free energy of transfer from a polar to a hydrophobic environment of the solvent. As a result, determining the ASA can assist in determining the binding strength. The total ASA for both of the complex's component proteins is divided by half to get the mean ASA.

For instance, the size calculations have helped to observe that three permanent heterocomplexes *viz.* Cathepsin D, reverse transcriptase, and hCG have comparatively larger interfaces for their molecular weights.

The shape of protein complexes is also analyzed since it is crucial while designing molecular models. Two protein subunits interact and create a protein combined with the two partially leveled planes or they could form a screw surface. This is determined by measuring exactly how far the interface residues have deviated from a plane which is called planarity. Planarity is evaluated using rms deviation which has varied results in the case of homo and heterocomplexes respectively. Electrostatic complementarity between the interfaces is another characteristic of the interaction between the proteins which is calculated using the Gap Index.

4.1.2.1.2 Complementarity between surfaces

A supplementary property used for Protein–protein docking methods is electrostatic complementarity between interfaces. It is usually calculated using the Gap Index which is the ratio of gap volume between molecules and the interface ASA. A mean Gap Index calculation of complexes leads to the observation that the interacting surfaces in the homodimers, enzyme-inhibitor complexes, and the permanent heterocomplexes have the highest complementarity while antigen and antibody complexes and non-obligatory heterocomplexes have the lowest complementarity.

4.1.2.1.3 Residue interface propensities

It is crucial to compare the respective importance of the various amino acid residues in the interfaces of complexes to provide a general idea of the hydrophobicity, which in turn can only be determined when the information is related to the distribution of residues present on the protein surface as a whole and has been talked in further sections. As a result, the residue propensities for each amino acid (AAi) that pitches into the interface as a fraction of ASA were examined, and the fraction of ASA that AAi pitches into the entire surface are compared. A propensity value larger than one indicates that a specific amino acid residue is found habitually in that interface instead of the surface. Another finding is that in heterocomplex surfaces, decreased propensities for hydrophobic residues are frequently offset by higher propensities for polar residues.

4.1.2.1.4 Hydrophobicity including Hydrogen bonding

The most frequent assumption is that proteins connect through hydrophobic patches on their surface, but another common association is polar contacts between subunits, therefore it's crucial to figure out how much each of these factors contributes to determining the driving force for complexation. For various types of complexes, a mean hydrophobicity value is determined, and it was discovered that the interface had an intermediate hydrophobicity between the interior and exterior hydrophobicity. Heterocomplex interfaces were also found to be less hydrophobic than homodimer interfaces. The roles of the various sorts of complexes have an impact on this. On the one hand, the hydrophobic surfaces of homodimers are permanently immersed within the Protein–protein complexes, whereas heterocomplexes frequently occur as monomers, so their interfaces cannot match the hydrophobicity of that of homodimers, otherwise, a good portion of the surface would become energetically favorable.

4.1.2.1.5 Segmentation and secondary structure

Another important characteristic is the quantity of interrupted fragments of the peptide chain constituting the interface as it helps determine the potential of the protein chains or even small molecules to copy about fifty percent of the interaction.

The discontinuous property of interfaces with respect to the amino acid sequences is determined using the calculation of the mean number of segments in the interfaces for each type of complex. Following this, interface amino acid residues split up with greater than 5 amino acids are put into contrasting sections, such that the number of segments are varied from 1-11.

4.1.2.1.6 Conformational changes on complex formation

Proteins have been observed to change their conformation as a result of complex formation and a few proteins have now even been structurally analyzed before and after the complexation using either X-crystallography or NMR.

Several levels of conformational changes are monitored using no change, movement of side chains alone, movement of segments having the main chain, and domain movements. For instance, a wide range of changes have been observed while studying the binding of antigen-antibody complexes, or during substrate binding. In general, it has been expected that both rigid as well as flexible docking can happen in different cases of complexation, but this also leads to changes in energy requirements to reduce flexibility. There are different types of interactions that are based on stability, composition, and affinity which will be discussed further.

4.1.2.2 PPI types
4.1.2.2.1 Homo oligomeric and Hetero oligomeric

As the name suggests, if the interacting proteins are identical then the interaction is called homo-oligomer interaction and on the other hand, if the interactions are taking place between non identical chains, then they form a hetero-oligomer. Homo-oligomers are symmetrical and thus are more stable, but the stability of hetero-oligomers varies. Homodimers are generally observed in the oligomeric form and cannot be separated into monomers (Keskin *et al.* 2016).

4.1.2.2.2 Obligate and non obligate complexes

If the interaction is obligate or non-obligate, it is determined by the protein's stability and affinity in both complex and monomeric states. Non-obligatory interactions involve proteins that are individually stable, but obligate interactions entail monomeric units that are unstable on their own (Levy and Teichmann, 2013). Obligate interactions are also known as two-state folders

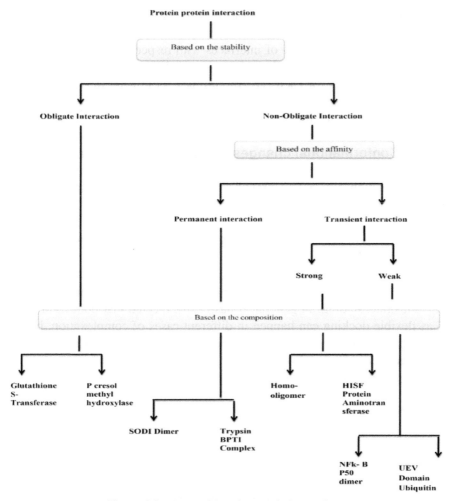

Figure 4.1   Types of Protein–protein interactions

(folding and binding happen simultaneously), whereas non-obligate inter-actions are three-state folders (proteins fold followed by coalescence and complex formation) (Keskin *et al.*, 2016).

4.1.2.2.3 Transient and permanent complexes

Based on the list of complex proteins to date, protein interactions are divided into two types. Interactions that remain stable and interact permanently are called permanent interactions whereas the interactions that associate and dissociate temporarily are called transient interactions (Jones and Thornton, 1996). This type of interaction can be seen in signaling and regulatory pathways because it allows cells to respond to the stimuli quickly and helps in a relay of the signal when required. Examples are inhibition of proteases, binding of hormones with their receptor, etc. Obligate interactions are generally ever-lasting whereas the non-obligate interactions are either permanent (Antibody-Antigen interaction) or temporary (large stable supramolecular systems). The stability and affinity of the complex are determined by binding free energy (Özlem Keskin et al., 2016).

4.1.2.2.4 Disordered to ordered complexes

Because disordered proteins have certain unstructured sections (big or small) whose amino acid composition cannot generate a stable folded shape, they do not form characterized Protein–protein interactions (Dunker et al., 2001) Like the tails of histone proteins, eukaryotic proteins are mainly disordered. Big dis-arranged fragments can fold at the same time while binding to their target in the case of large disordered segments (Özlem Keskin et al., 2016).

4.1.2.3 PPI methods classification

Interaction between proteins can be experimentally determined both at a small scale and large scale with two technologies (binary and co-complex) that produce varied data.

The Binary method measures head-on solid interaction among the proteins. The most common method used is Y2H (Yeast two hybrid). The co-complex method computes the physical distance linking the proteins without determining the pairwise protein partners. Tandem affinity purification coupled with mass spectrometry is the most commonly used method in the case of co-complex methodology. The results obtained from both of the methods are quite different from each other. We cannot directly assign a binary operation on the results that are obtained from a co-complex instead a model is used which converts the batch observations to matched set interaction. One of the models, that is commonly used is the Spoke model as it gives less amount of false-positive results (Rivas and Fontanillo, 2010).

4.2 *In Silico* Methods

The invention of useful tools for analyzing protein interactions (PPIs) between specific proteins that exist in various coalescence was made possible through many *in vivo* and *in vitro* methods. But, due to the lack of available

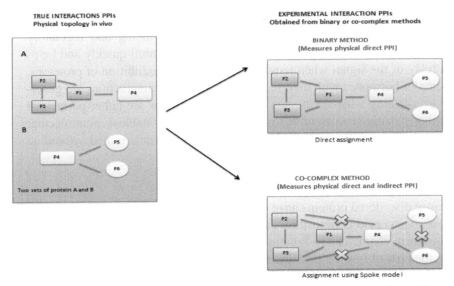

Figure 4.2   Binary and Co-complex method for Protein–protein interaction (adapted from Rivas and Fontanillo, 2010). Here, X represents the interaction that does not take place (i.e., they are false positive results)

PPIs, the data provided by these methods may not be reliable. It is preferable to create techniques that speculate the whole list of probable inter-linkage among proteins of interest, to comprehend the entirety of likely interactions. To support the interactions discovered by the experimental technique, a range of *in silico* methods have been developed. In the coming sections, the various computational methods for *in silico* prediction have been discussed.

4.2.1 Sequence Based Approaches

PPI predictions have been made using already present data related to protein interactions combined with details on homology among the sequences. This method relies on the idea that an interaction observed in a certain species can be extrapolated to other species (Rao *et al.*, 2014).

Hosur *et al.* (2011) used a threading-based approach by feeding sequences to develop a novel method to foresee protein interactions. The iWARP (Interface Weighted RAPtor) technique, which compiles a unique linear programming approach for interface arrangement, also has an enhancing classification of inter-linkage projection, and speculates if two proteins interact (Rao *et al.*, 2014).

Guilherme Valente et al. proposed a novel technique for identifying protein pairs as interacting or non-interacting proteins based on primary sequence information termed the Universal *In Silico* Predictor of Protein–Protein Interactions (UNISPPI) (Valente *et al.*, 2013).

4.2.1.1 Ortholog based sequence approach

Sequence-based prediction employs a similarity-based method of elucidation, shifting from a known protein with a well-defined function to the sequence of interest. Pairwise local sequence alignment is based on the query protein's similarity index in an established protein database. Several proteins found in the organism under inquiry may mimic proteins found in other organisms that are involved in structure formation (Lee *et al.*, 2008)

The initial stage in the analytical procedure is to compare a tagged protein to known proteins from different organisms. It is often presumed that our protein of interest has activity or features similar to that of the known function protein in another species if the sequence of the labeled protein matches that of a known function protein in another species (Rao *et al.*, 2014).

This was observed for a majority of components of protein complexes. Ortholog and paralog ideas should be used when transferring a function out of a recognized protein to a protein that has not been characterized. Orthologs are genes that have progressed from a gene of common progenitors in distinct animals through speciation. Paralogs, on the other hand, are genes that are associated with a genome due to duplication. Orthologs will, in general, preserve their functionality as they evolve, but paralogs may acquire new functionalities (Rao *et al.*, 2014).

As a result, if two proteins–X and Y–interact with one another, orthologs of X and Y in a new group are expected to inter-relate as well.

4.2.1.2 Domain pairs-based sequence approach

A domain is a protein's distinct, compact, and stable structural unit that folds independently of other subdomains. Domains, on the other hand, are typically thought of as discrete regions of the protein sequence that have evolved to be highly conserved. Protein domains have separate structural and functional components, and have a crucial part in the construction of prediction of protein structure (Rao *et al.*, 2014).

Protein domains are extensively employed in fundamental research and drug discovery based on structure. Domains should also be crucial in protein interactions because they are a part of intermolecular interactions. Various researches have demonstrated that domain-domain interactions (DDIs) from various trials are more resolute as compared to their corresponding PPIs. As a result, predicting Protein–protein interactions using domains is exceedingly accurate (Wojcik and Schächter, 2001).

4.2.1.3 Statistical sequence-based approaches

4.2.1.3.1 Mirror tree method

The Mirror Tree Method was created by Florencio Pazos and Alfonso Valencia. It evaluates the tree structure of phylogenetic trees and determines the similarities among such trees. These data are used to estimate mirror tree method-based PPIs by analyzing evolutionary distances between connected protein families' sequences. (Pazos and Valencia, 2001).

These distances were calculated using the McLachlan matrix's mean amino acid analogy. The distance matrices were used to calculate tree similarity. The Mirror Tree Method does not need the production of phylogenetic trees; instead, includes the analysis of the basal distance matrices, making it independent of any tree-generation method (Pazos and Valencia, 2001).

While the mirror tree method may not need whole sequenced genome structures, it does require the appearance of orthologous proteins among the list of the species studied. This leads to a step where, as the number of genomes grows, only a small number of proteins may be considered. Furthermore, the technique is limited to circumstances in which a minimum of 11 sequences from related species were collected for all proteins in question. The lower limit was calculated through trial and error as a settlement between being reasonably minuscule to give many cases and large enough to allow the matrices to carry enough data. To increase the number of possible interactions, the approach can be improved by assembling sequence data from an appreciable number of samples of genomes. Furthermore, because distance matrices don't completely represent connected dendrograms, comparing distance matrices rather to real phylogenetic trees may introduce problems (Shatnawi, 2015).

4.2.1.3.2 PIPE

Protein–protein Interaction Prediction Engine was developed by Pitre et al. to predict the probability of interlinkages between protein structures of S. cerevisiae based on protein polypeptide sequence information. The PIPE algorithm is constructed on the fact that protein interlinkages occur through a definite amount of concise amino acid sequences found in an established database of proteins (Pitre *et al.*, 2006)

These are usually smaller than traditional domains and can be found in a variety of proteins throughout the cell. The recurrence of these small polypeptides among known interacting protein pairs is used by PIPE to evaluate the likelihood of a PPI (Pitre *et al.*, 2006).

To assess if proteins A and B interact, the two query proteins are compared to a database of known interacting protein pairings. For each annotated interaction pair (X, Y), PIPE compares the amino acid residues in protein A to those in X and protein B to those in Y, then counts the probability of a protein A finding a match in X and a comparable result in protein B finding a match in Y. These matches are counted and totaled in a 2D matrix. A positive protein interaction is anticipated when the reoccurrence count in specific matrix cells surpasses a preset threshold value. When there isn't enough structural information, it can detect interactions between protein combinations (Shatnawi, 2015).

The following are PIPE's restrictions. PIPE is a computationally demanding program that takes hours to run for each pair of proteins because it surveys the library multiple times. Next, because it revealed a substantial number of false positives, PIPE reveals a deficit in discovering novel connections among genome-wide large-scale datasets. Third, PIPE was tested using unreliable data on interactions derived from a variety of methods, each of which had a limited accuracy (Shatnawi, 2015).

Pitre et al. later created PIPE2, a better version of PIPE with a precision of 0.999. The latest version encodes AA sequences in the format of zero's and one's, which allows it to explore the similarity matrix more quickly. However, despite its high specificity, PIPE2 has a high amount of false positives, with a sensitivity of only 0.146. PIPE2 has a severe flaw in that it depends solely on existing protein pairings to identify recurring short polypeptide sequences, and it will be unsuccessful if there isn't enough data. Because it does not account for gaps within short polypeptide sequences, PIPE2 is less successful for motifs that span discontinuous source sequences (Pitre *et al.*, 2008).

4.2.1.3.3 Co-evolutionary divergence

In human proteins, Liu et al. developed co-evolution based analytical techniques. The writers based their definition of co-evolutionary divergence (CD) on two assumptions. First, PPI pairs may have identical substitution rates. Second, protein interactions have a high probability to be conserved among species. CD is the total measure of the difference in substitution rates in the case of our proteins of interest. CD can be used to predict PPIs since interacting protein pairs are predicted to have lower CD values than non-interacting protein pairs (Hsin *et al.*, 2013).

Co-evolutionary evidence from protein interactions involving several organisms is incorporated into the CD technique. It is faster as compared to additional alignment methods, like the mirror tree method, because it doesn't use multiple alignments. The method isn't limited to orthologous proteins from different species. Increasing the sample space improves the precision of the co-evolutionary divergence technique. Even though the technique was successful in predicting the probability of a protein interaction for a given sample space, it was unable to anticipate key elements of the interaction, such as the participating residues at surfaces (Shatnawi, 2015). Protein interactions, on the other hand, can be studied using a machine learning sequence-based technique.

4.2.1.4 Machine learning sequence-based approaches
Protein interactions are vital for cellular functions. Thus, it becomes crucial to be able to identify as well as visualize interactions between not just two proteins but also more than two proteins. The following include the different types of machine learning approaches based on sequence analysis.

4.2.1.4.1 Auto-covariance
Seven physiochemical properties were used to represent amino acids (AA) *viz* hydrophobicity, hydrophilicity, polarity, side chains' volumes, polarizability, the surface area available to solvent, and the total charge index of AA side chains (Guo *et al.*, 2008).

Auto covariance (AC) is associated with the binding interactions that occur between aa residues a particular length apart in a sequence. Thus, based on the determination of auto-covariance, the physiochemical properties of AA were studied, followed by the characterization of an entire protein sequence, thereby containing the information regarding the interactions occurring between every aa as well as the rest of the peripheral residue present in the sequence.

Similarly, a Support Vector Machines technique was developed based on the values of AC. All this information related to each and every AA helped in the formation of the PPI network. The main advantage of AC is the binding information of aa residues even in long ranges.

4.2.1.4.2 Pairwise similarity
It is created on the concept of the percentage similarity of the primary sequence of a pair of proteins. A window is made to slide over the connected

protein sequences of interest, while the E-value is determined based on the Smith-Waterman algorithm. The subsequent matrix formed is used in a combination with the SVM technique (Zaki *et al.*, 2009).

The method helps in providing a level of similarity between two proteins, which may further indicate a certain level of homology and has now become a foundational tool in many of the PPI networks and computational biology. The combined use of SVM with pairwise similarity aids in other areas as well, such as the determination of potential sites as drug targets, gene ontology, and inter-domain linker regions (Shatnawi, 2015).

4.2.1.4.3 Amino acid composition

It has long been agreed upon that determination of the entire set of possible and actual protein interactions is a must for studying how cells function, however, there are still some false negatives that later on go unrecognized. A mathematical prediction of the composition of a protein proves to be an inexpensive as well as an efficient method for the highly likely protein interactions possible. Protein domains naturally become the option in determining the static properties of a protein structure, since they have the added advantage of evolutionary conserved domains (Roy et al., 2009).

Hence, the basic idea includes establishing the residue composition of a protein molecule that then helps develop monomer traits and the information is then segmented into minimum-sized bits, thereby establishing binary features. This approach has been practically utilized in the analysis of model organisms namely yeast, worm, and flies. Determining the amino acid composition is relatively economical, can be used for all polypeptides, and can even prove useful in case of a lack of information available related to domain organization (Shatnawi, 2015).

4.2.1.4.4 Amino acid triad

Although the earlier discussed methods may be efficient, they sometimes prove to be a bit expensive to obtain (Yu *et al.*, 2010). Yu et al. gave a method that uses probability to estimate the importance of triads in increasing the impact of amino acids organization in nature. AA triads are three continuous residues in the protein sequence and are considered one unit. These triads are then scanned one at a time, wherein every single entity is called the occurrence vector i.e., O.

Later on, a significance vector, i.e., S is given in order to denote a sequence of proteins through the possibility of actually getting lesser triads as compared theoretically. Information was then put in the PPI tool for the prediction of possible interactions. It used the data from HPRD (Human Protein Reference Database). The results included a positive-negative ratio which in turn indicated sensitivity, accuracy, sensitivity, and F value (Shatnawi, 2015).

4.2.1.4.5 UNISPPI

The approach involves consideration of frequency and composition of the physiochemical features of 20 AA in order to construct a decision tree PPI. It is a probability-based technique and has the following advantages such as the convenient exclusion of the non- PPIs, the vast number of PPIs and non-PPIs known is so much more than found in other techniques, and the highly reduced costs for carrying out the technique (Valente *et al.*, 2013).

With the analysis of more than two species at a time, the UNISPPI (Universal *In Silico* Predictor of Protein–Protein Interactions) gave a distinct range for PPI as well as non-PPIs and also used the frequency occurrence of Asparagine, Cysteine, Isoleucine to differentiate between PPI and non-PPIs. The method has only a few drawbacks, one being that such decision trees usually have a side effect of overfitting (Shatnawi, 2015).

4.2.1.4.6 ETB viterbi

The Hidden Markov Model (HMM) is basically a probability-based method, applied in the description of the evolution of events that are affected by intrinsic factors but may not be detectable. The incident here is designated as a 'Symbol' while the internal ingredient affecting the event is termed as a 'state'. One practical application of HMM is observed in the process of gene prediction, pairwise and multiple alignments of sequences, predicting DNA sequencing errors, and many more areas of biology (Yoon, 2009).

However due to the hindrance of not being able to determine relations over long distances without preventing any mathematical errors, ipHMMs (interaction profile HMM) were found to lagging, but long-distance relations between residues could not be denied (Kern *et al.*, 2013).

Thus, the Early Traceback Viterbi (ETB-Viterbi) was an efficient replacement to calculate long-range distances which were based on the data acquired from the 3DID database along with numerous simulations, hence resulting in the discovery of the most probable sequence of hidden states but still proves to be a more expensive approach than HMM modeling (Shatnawi, 2015). Another different approach involves analysis based on interactions influenced by respective protein structures.

4.2.2 Structure Based Approaches

The structure-based method works by predicting Protein–protein interaction when there is a structural similarity between the proteins. Three-dimensional structural characteristics are used in structure-based PPI prediction algorithms. As a result, if two proteins X and Y interact, there is a possibility that X' and Y' are structurally alike with X and Y; this implies that proteins X' and Y' can interact as well (Rao *et al.*, 2014) (Shatnawi, 2015).

However, because the structure of most proteins is unknown, the initial stage in the procedure is to postulate the protein organization using the known sequence of the protein. The structure prediction can be accomplished in a variety of ways. The PDB database provides researchers with essential mechanisms and data sources for constructing the protein layout (Rao *et al.*, 2014).

Template-based, statistical, and machine learning-based PPI prediction approaches are the three types of structure-based PPI prediction methods (Shatnawi, 2015).

4.2.2.1 Template structure-based approaches

When close templates are obtained, template-based approaches can obtain extremely efficient prediction accuracy, but in case there is a less similarity index between the template and the protein of interest, the accuracy drops dramatically.

4.2.2.1.1 PRISM

PRISM is a template-based PPI prediction approach structured on the knowledge of the interaction between the surfaces of the complex protein molecules. It was created by Tuncbag et al. (Shatnawi, 2015)

Structural alignment compares either side of the concerned interface with the individual surface of the two monomers under consideration. The two proteins of interest are predicted to bind through the surface structure of the template if the complementary sides of the template interface are identical to parts of the target surfaces. Naccess is used to retrieve target chains' interacting surface residues. Second, MultiProt is used to distinguish complementary chains of template interfaces and compare them structurally to each of the target surfaces. Finally, the results are filtered using predetermined benchmarks, followed by the creation of template interfaces and the design of a possible combination (Shatnawi, 2015).

Interactions are strengthened by Fibre-Dock and the solutions are ranked based on their energies. When a protein pair's computed energy is less than a threshold of -10 kcal/mol, the pair are said to interact.

PRISM has been used to predict PPIs in human apoptosis and a p53-related pathway, and it has helped researchers better understand the structural principles underpinning various types of signal transduction. When applied to a human apoptosis pathway made up of 57 proteins, PRISM had a precision of 0.231 (Ozbabacan *et al.*, 2012) (Tuncbag *et al.*, 2009).

4.2.2.1.2 PREPPI

Zhang et al. introduced PrePPI (Predicting Protein–Protein Interactions) as a structural alignment Protein–protein interaction predictor based on secondary structure data geometric correlations. For a pair of query proteins P and Q, representative structures for the constituent subunits (MP, MQ) are chosen from the PDB (Protein Data Bank) or other databases. If we could find any PDB structures that are structural neighbors of M_P and M_Q then it serves as a template for the interaction. The component subunits, M_P and M_Q, are superposed on their appropriate structural neighbors to create a model. A Bayesian Network is then used to calculate the likelihood of each model representing an actual interaction. Finally, the score which is derived from the structure and non-structure derived scores are combined into a naïve Bayes classifier (Zhang *et al.*, 2012) (Shatnawi, 2015).

4.2.2.2 Statistical structure-based approach

4.2.2.2.1 PID matrix score

Kim et al. introduced the Potentially Interacting Domain pair (PID) matrix as a domain-based PPI prediction algorithm. The matrix score is calculated using the probability of interaction between the domains (Kim et al., 2019).

The approach was built on the DIP (Database of Interacting Proteins), which comprises over 10,000 interacting protein pairings, the majority of which have been experimentally confirmed. InterPro is used to extract domain information. Sub-divisions of DIP data (positive datasets) and arbitrarily produced protein matches using the TrEMBL/SwissProt database were used for cross-validation (negative datasets). The approach had a sensitivity of 0.50 and a specificity of 0.98. The PID matrix can be used to map genome-wide interaction networks as well (Poluri *et al.*, 2021).

4.2.2.2.2 Pre SPI

Hans and his colleagues proposed a domain combination-based approach, in which all possible domain combinations are treated as fundamental components of protein interactions. The number of interacting protein pairings containing the domain combination pair, as well as the number of domain combinations that might be generated in each protein, can be used to compute the probability of domain interaction (Hans *et al.*, 2004).

The technique takes into account the likelihood of domain amalgamations showing up in all kinds of interactions. The interacting probabilities estimated using the interacting probability equation was used to rank different protein pairings. This method was first evaluated in the yeast system using the set of proteins that interact with each other, retrieved from the DIP database, and the non-interacting protein dataset which was created randomly.PDB was used to extract information about the domains of the proteins. PreSPI attained a sensitivity of 0.77 and a specificity of 0.95 (Shatnawi, 2015).

There are various drawbacks to PreSPI. First, relevant data regarding the other domain combinations are ignored by this method. Also, it presumes that each domain interaction is independent. Thirdly, because all possible domain combinations are investigated, the method needs high computational power and resources (Shatnawi, 2015).

4.2.2.2.3 Domain cohesion and coupling

Using intra-protein and inter-protein domain interactions, Jang et al. suggested a domain cohesion and coupling to the established protein interaction prediction approach. The technique seeks to pick the domains that are a part of protein interaction by calculating the likelihood of domains that cause interaction between proteins, regardless of the number of domains involved (Jang et al., 2004).

The coupling powers of all domain interaction pairs are stored in an interaction significance (IS) matrix, which is used to forecast PPIs. The approach had a sensitivity of 0.82 and a specificity of 0.83 on S. cerevisiae proteins. The proteins' domain information was extracted using Pfam, a database of protein domain families (Shatnawi, 2015).

4.2.2.2.4 MEGADOCK

MEGADOCK is a Protein–protein docking software tool developed by Ohue et al., that uses the real Pairwise Shape Complementarity (rPSC) score. Rigid-body docking calculations were performed for every conceivable protein pair using a simplified energy function that took form complementarities, electrostatics, and hydrophobic interactions into account (Ohue *et al.*, 2014). Using this method, For each pair of proteins, a collection of high-scoring docking complexes was obtained. The docking data were then re-ranked based on

ZRANK energy ratings for a more advanced binding energy computation. A standardized score (Z-score) was calculated and used to analyze probable interactions. Structure differences were used to rule out potential complexes that didn't have any other high-scoring interactions nearby. As a result, only binding pairs having at least one populated area of high-scoring structures were taken into consideration. The precision of MEGADOCK is 0.4 (Ohue *et al.*, 2014).

$$Z(G_k) = \frac{f_{real}(G_k) - \bar{f}_{rand}(G_k)}{std(f_{rand}(G_k))}$$

$$\Delta(G_k) = \frac{f_{real}(G_k) - \bar{f}_{rand}(G_k)}{f_{real}(G_k) + \bar{f}_{rand}(G_k) + \epsilon}$$

f real(Gk)- frequency of the subgraph Gk into the real network, while f rand(GkÃđ) and std(f rand(Gk)) are the averages and the standard deviation of the frequency of Gk in the set of random networks, respectively (Ciriello and Guerra, 2008).

MEGADOCK 4.0, an ultra–high–performance docking program that makes considerable use of supercomputers equipped with GPUs, is now being deployed (Ohue *et al.*, 2014)

One of the drawbacks of this method is that it can produce false-positives in circumstances where no similar structures can be found in known complex structure databases.

4.2.2.2.5 MetaApproach

Ohue et al. suggested a method for predicting PPIs that combined template-based and docking methods. PRISM as well as MEGADOCK is used as docking mechanisms relying on template similarity. If both indicate a possible interaction, its accuracy is confirmed. The method acquired a precision of 0.333 when applied to the human apoptotic signaling system, which is higher than that achieved by separate methods (0.231 for PRISM and 0.145 for MEGADOCK). In the discipline of protein tertiary structure prediction, Meta approaches have already been employed, and rigorous experiments have shown that Meta predictors outperform individual methods (Shatnawi, 2015).

Protein domain prediction and prediction of disordered areas in proteins have also yielded positive results using the Meta technique. Although this strategy may eliminate some true positives, the remaining predicted pairs are projected to have a greater level of dependability due to the agreement between two prediction methods with different characteristics (Ishida and Kinoshita, 2008). Machine learning structure-based approaches are now being considered as an alternative to Protein–protein analysis research.

4.2.2.3 Machine learning structure-based approaches

4.2.2.3.1 Random forest

The method was based on the existence of particular domains that occur in both proteins. With the information acquired from the Pfam database, a structure (domain) based approach was introduced and was named the Random Forest PPI predictor.

Domains provide a structural basis for protein analysis as they are the structural and functional units of almost all proteins and remain conserved through generations and help in the identification of basic protein structures as well as functions, a wide assumption being, domains are the connecting links in protein interactions (Chen *et al.*, 2005).

The Presence of a domain in both the proteins has a feature value of 2, the presence of a domain in only one of two proteins of interest has a value of 1 while the absence of a domain in both the proteins gets the score 0.

The Random Forest PPI predictor has had an immense effect on the recognition of hotspots from the general majority population of non-hotspots, by a generation of a set of forest trees through oversampling thereby giving rapid and precise results (Zhang et al., 2019), even though it might not be able to prove useful for cases where not much information is available for domains (Shatnawi, 2015).

4.2.2.3.2 Struct2Net

In the case of large interactomes, computational techniques aid in the precise determination of protein interactions. Experimental methods might be inclined to show errors that are then overcome by structure-based methods. Struc2Net involves the answering of yes or no to enormous functional information. Even though sequence-based domain methods are known for the identification of domains that are a part of protein interaction, structure-based methods have an edge in insight, provision of functional annotations, and increased performance (Singh et al., 2006).

The method revolves around scanning each pair of protein sequences with every possible structure from the information gathered from the Protein Data Bank. After going through all protein sequences, Struc2Net gives the best possible protein interaction. (Shatnawi, 2015).

4.2.3 Gene Neighbourhood

The basic information for searching protein interactions has been increasing with the continuous progression in the sequencing of genomes of not only model organisms but many more organisms and the data is still growing. An established pattern observed by scientists is that functionally related proteins have the tendency to cluster together closely in the genomes of organisms, the clusters called mRNA. The method is based on the very concept of conservation of such clusters across the genomes of closely related organisms and was experimentally confirmed. The level of similarity indicates the degree of closeness (Rao et al., 2013).

Gene neighborhood analysis of proteins unknown using the physico-chemical properties of amino acids is popular method during drug discovery. Two such methods, FunPred1.1 and FunPred1.2 use the concept of gene neighborhood where FunPred1.1 is based on the whole neighborhood graph of unknown proteins while FunPred1.2 has a higher efficiency (Saha et al., 2017).

4.2.4 Gene Fusion

Popularly known as the Rosetta Stone Method, it is constructed on the possibility that some proteins consisting of only one domain in one organism tend to fuse and result in a multidomain protein when present in other species. The ability of fusion denotes the prospect of functional association of such proteins, which might fuse to produce a stable Protein–protein interaction (Rao et al., 2013).

The method has seen applications in predicting various disease genes from PPI networks. The algorithm focuses on the analysis of clinical single sample-based PPI networks (dgCSN) that does not rely on the stats of PPIs which might change during the life cycle of an organism (Luo et al., 2018).

4.2.5 *In Silico* Two-hybrid (I2h)

Based on the presupposition that the proteins involved during binding must go through the process of evolution to ensure the continued proper functioning of the protein. The assumption can be rephrased as the damage caused due to the likelihood of a change in the vital amino acids in one protein, must be controlled by the simultaneous change in the other protein to ensure that there is no overall impact on the Protein–protein interaction (Rao et al., 2013).

I2h is a strategy built on the analysis of sequence interdependence in the remote positions, although weak, but is a crucial factor in the prediction of inter-residue exposure. The procedure follows the search for multiple sequence alignments with a well-defined signal which can be made possible on the basis of co-evolutionary mutations between the protein sequences of interest (Pazos *et al.*, 2002).

4.2.6 Phylogenetic Tree

A phylogenetic tree provides the evolutionary history of a protein. Protein interactions are determined using the earlier discussed mirror tree method with the assumption that proteins involved in the interaction must show some degree of similarity when a phylogenetic tree is studied. This is because of the co-evolution of proteins that are a part of the interaction. Such a co-evolution can be represented by the usage of distance matrices of the protein dendrograms in an interaction complex (Rao et al., 2013).

High scores of similarities in the resulting phylogenetic trees and corresponding distance matrices denote a high prospect of protein relationships. Hence, based on the mirror tree approach, the likelihood of co-evolution between the two proteins of interest can be used to infer the chances of an actual relation (Malik et al., 2017).

4.2.7 Phylogenetic Profile

The principle of co-evolution of two proteins throughout the generations can be explained as the pressure existing on the proteins to be passed on together during the process of evolution, which may further imply that the corresponding orthologs in their respective genome structures may either be retained or discarded. This is the basis of working in a phylogenetic profile (Rao et al., 2013).

The information gathered from the enormous databases and distance relationships forms the backbone of studying phylogenetic profiles. One recent example includes the study of PPIs involved in the Plasmodium falciparum mechanism of attack on the host and the subsequent parasitosis symptoms. The approach assisted in prognosticating the possibility during the exposure to the parasite that is to say host-host, host erythrocyte, etc PPIs (Sterra et al., 2018).

However, there might be the points like noise due to gene duplication or loss of gene functions, dependence on entire genome sequences, and lack of detection of protein interactions which must be considered when studying PPIs using phylogenetic profiling.

4.2.8 Gene Expression

The procedure of the evaluation of the extent to which a certain gene may be transcribed and translated inside a cell, within an organism is called gene expression. Utilizing the algorithms based on clustering, various expression genes can be arranged in conjunction depending on their respective levels of expression while the different sets of physiological conditions and resultant gene expression reflect on the functional interdependence of the genes under study. Such an analysis has established a relation between the gene expression and Protein–protein interactions (Rao et al., 2013).

The concept of gene expression has been applied in the study of various cancer types. The use of the cDNA microarray method aids in determining the transcription level of a cell at an instantaneous time during the life cycle, thus resulting in the analysis of a genome wide expression fashion of the cells that in return provide an understanding of the internal molecular level processes that are a part of cancer functioning (Verma *et al.*, 2019). All popular approaches have been compiled in the table given below for efficient understanding.

Method	Principle
Structure-based approaches	If two proteins have a similar structure, structure-based techniques can predict protein-protein interaction (primary, secondary, or tertiary)
Ortholog-based sequence approach	Using the pairwise local sequence technique, an ortholog-based sequencing approach based on the homologous nature of the query protein in annotated protein databases was developed.
Domain-pairs-based sequence approach	Based on domain-domain interactions, a domain-pairs-based approach predicts protein interactions.
Phylogenetic profile	If two proteins have the same phylogenetic profile, the phylogenetic profile predicts their interaction.
In silico 2 hybrid (I2H)	The I2H technique is based on the idea that interacting proteins should coevolve in order for the protein function to remain consistent.
Gene neighborhood	If the gene neighbourhood is conserved across many genomes, functional connection between the proteins encoded by the associated genes is a possibility.
Gene expression	Proteins from genes in similar expression-profiling clusters are more likely to interact with each other than proteins from genes in distinct clusters, according to gene expression.
Gene fusion	Gene fusion, often known as the Rosetta stone method, is based on the idea that some single-domain proteins in one organism can combine to generate multi domain proteins in other animals.
Phylogenetic tree	Based on the protein's evolutionary history, the phylogenetic tree technique predicts the protein-protein interaction.

4.3 PPI Networks and Databases

It often becomes tricky when the need arises to study Protein–protein inter-actions of more than two proteins or even a case in which not much is known about the structural organization of the protein. Such interactions are analyzed *via* probing and mathematical methods that have been created in order to minutely recognize protein interactions (Hans *et al.*, 2004).

A network of linking proteins is prepared using a graph, in which the proteins are shown using nodes and the interactions among proteins are illustrated by edges. It also includes the presence of sub-graphs due to different properties related to the main graph. Thus, PPI networks prove to be especially useful in the case of recurrence of some particular modules and these may have a certain meaning to them, for instance, a conserved protein unit (Ciriello and Guerra, 2008).

There are two definitions revolving around motifs. They are based on the frequency and statistical importance respectively. Where on one hand, a motif is described as a subgraph that appears more than a certain number of times if the frequency is the basis, while on the other hand, a motif could be also defined as a subgraph that occurs a greater number of times than the expected chance, statistically speaking. Though it seems six and two threes, they are different definitions altogether (Ciriello and Guerra, 2008). Motif discovery is made possible through algorithms including Exact algorithms and Approximate algorithms, depending on static and temporal PPI networks (Jazayeri and Yang, 2020). The major steps included are as follows:

4.3.1 Creation of the PPI Networks

4.3.1.1 Choice of databases and data selection

The primary choice of data and corresponding databases are popularly done using the list of databases present in 'Pathguide', the Pathway Resource List, a metadatabase (Ciriello and Guerra, 2008). It contains not only protein sequence information, and Protein–protein interaction databases but also metabolic, signaling, and gene regulation pathway databases (Bader et al., 2006). As important as the selection of databases, the next crucial step is the visualization of plausible Protein–protein interactions in all conformations.

4.3.1.2 Visualising PPI network

Visualization of databases can be performed using open-source Cytoscape wherein each one of the networks was designated by nodes and interactions by edges. The advantage of Cytoscape over STRING is flexibility with respect to network analysis, import, and visualization of additional data (Doncheva *et al.*, 2018). The nodes are linked to the Pathguide ID. The files could be accessed in a CYS file, i.e., the format used in Cytoscape. It also includes three types of shapes for edges namely arrow, diamond, and straight line, besides the added feature of color-coded databases (Ciriello and Guerra, 2008). The next section sheds light on the various databases that have been compiled based on the information gathered based on analysis and visualization of different protein interactions.

4.3.2 Different Databases

With the ever-increasing information regarding PPIs in living systems, systematic organization of the computational and experimental data becomes the sine qua non of studying protein functions in bioinformatics.

Three distinct data sharing patterns can be seen within PPI source datasets.

1. Standalone databases that do not actively share data (HPRD, BIND, BioGRID)
2. Generalist databases that participate in the IMEx agreement (DIP, MINT, IntAct)
3. Topical databases that merge meta mining of different databases with their domestic curated material (MatrixDB, DroID, MPIDB, InnateDB)

Topical databases appeal to scholars because of their combination of in-house curation and substantial meta mining. Only data curated in compliance with the IMEx curation guidelines are shared in IMEx databases. It implies that IMEx exchange needs to be augmented with a meta mining procedure. Unless it is done, both the non-sharing as well as the IMEx databases are in most probability have a substantial number of entries unique to the database. Researchers may find the use of meta mining databases to be more convenient than the time-consuming searches in the generalist source databases (Klingström and Plewczynski, 2011).

4.3.2.1 Interaction database

The Protein–Protein Interaction Databases are targeted towards specific organisms like Drosophila melanogaster (DroID), microorganisms (MPIDB), or targeted towards specific regions like extracellular matrices (MatrixDB) or in some health conditions like our immune system (InnateDB) which comprises the topical databases. Interaction databases appeal to scholars interested in the database's topical area because of the amalgamation of comprehensive data mining and in-house curation (Klingström and Plewczynski, 2011).

The HPRD and Flybase databases were introduced by BioGRID under its datasets in 2006. The BioGRID (Biological General Repository for Interaction Datasets, thebiogrid.org) is a free resource that comprises hand-curated protein and genetic interactions from a wide range of organisms, including yeast, worm, fly, mouse, and humans. The 1.933 million curated relationships in BioGRID can be used to build complex networks to facilitate biological research, particularly in genetics. BioGRID can also identify post-translational modifications and protein or gene interactions with bioactive small compounds, including numerous well-known medicines. (Oughtred *et al.*, 2021)

Data from other databases is not incorporated into BIND, DIP, HPRD, IntAct, or MINT. Currently, there is little overlap across most source datasets. Information from a restricted number of scientific works is acquired by every literature curating database thus only possibility is that of shared interactions when a single journal is curated by two different databases or a single protein interaction is analyzed in reports in different journals. IMEx consortium includes all other important databases that curate literature, except for HPRD. The initial step in this collaboration is a collaborative effort in literature curation, which leads to steady growth in shared data in databases. The IMEx members split the literature sources for curation, which are subsequently curated in an IMEx defined manner and made available to all IMEx members (Klingström and Plewczynski, 2011). For the purpose of extraction of data from entire databases, meta mining databases take up the next slot.

4.3.2.2 Metamining databases

The target of compiling disparate root data sources to a unique and exhaustive meta-database is achieved by the popular meta mining databases such as UniH, MiMI, and APID.

Another method of metamining interaction databases is to incorporate interaction networks into reaction pathways. ConsensusPath and Pathway-Commons describe their interaction networks in this way. The merging of information from pathway databases and data from interaction databases results in complex pathways (Klingström and Plewczy, 2011)

Meta-databases are intriguing sources of interaction data because they allow you to include many interaction databases in one search. It's vital to remember that every meta-database has its potential horizon, and as a result, many meta-databases have selected portions of the root databases.

Each meta mining database owns a unification method, which has its own set of strengths and drawbacks. The coverage of the various source databases is the most easily assessed component (Klingström and Plewczy, 2011).

The Amount of database covered is calculated based on the fractions of PPIs that are introduced in the metamining databases. We want to emphasize that coverage is not a measure of database quality because each meta-database has a unique range, that might not exactly match that of the source databases.

PINA is a network analysis tool with information from MINT, IntAct, DIP, and BioGRID already built-in. PINA also allows users to establish custom meta-databases by uploading individual PPI information. PINA is useful for researchers that want to undertake protein interaction studies on certain pathways or develop new ones using the information present in databases.

DASMI, which combines protein and domain interaction information, is another option. Instead of keeping all data in a single database, this version of the Distributed Annotation System queries all of the databases that are included and aggregate the information when the user requires it. Because a large amount of data must be transferred for each submission, decentralized techniques are less appropriate to wide-ranging research, such as GWAS, than meta mining databases (Klingström and Plewczy, 2011). Predictive databases, on the other hand, are directed at the most conceivable interactions between two proteins that can be predicted with the greatest accuracy.

4.3.2.3 Predictive interaction databases

HAPPI, STRING, STITCH, and Scansite are the four primary predictive PPI datasets. To determine their expected interactions, HAPPI and STITCH rely upon the STRING database. The fact that STITCH also makes predictions about Protein–chemical interactions give it a unique edge (Klingström and Plewczy, 2011).

STRING integrates existing information present in interaction databases like IntAct MINT, BIND, DIP, HPRD, BioGRID, and with pathway inter-connections from EcoCyc, PID, KEGG, and Reactome. It achieves this by collecting and scoring evidence from many sources. It is supplemented by inter-connections anticipated by algorithms created particularly for STRING. The STRING database aims to integrate all known and expected protein interactions, including both physical and functional interactions, together in one place (Klingström and Plewczynski, 2011).

HAPPI extracts human interactions from STRING and compares them to data from major databases such as BIND, HPRD, KEGG, MINT, and OPHID. All sources have their accuracy levels, and similar results of the interactions across many databases increase its authenticity, resulting in a database of about 138, 000 active supreme PPIs, often known as 3, 4, or 5 star-quality interactions in HAPPI terminology (Klingström and Plewczynski, 2011).

The Scansite database is built by comparing 1D sequences to recognized patterns. Small polypeptide sequences that might be recognized by signaling domains, which are prone to reactions by Serine, Tyrosine, or Threonine kinases, or arbitrate particular interactions with protein or phospholipid ligands, can be observed using Scansite (Klingström and Plewczynski, 2011). Yet another database focuses the light entirely on the metabolic interactions taking place in various organisms and provides a link between the vast number of pathways.

4.3.2.4 Pathway database

By constructing pathways to describe biological processes, pathway databases give PPIs a biological meaning. Some databases, like KEGG, track protein interactions directly, while others, like Reactome, use the approach of biochemical transition and transportation instead of protein interactions (Klingström and Plewczynski, 2011).

KEGG is a large-scale molecular dataset for biological interpretation. It collects experimental data on high-level functions of cells and organisms expressed as KEGG molecular systems, such as KEGG pathway maps, BRITE hierarchies, and KEGG modules. A group of genes in the genome that code for proteins, for example, can be made into KEGG molecular networks, allowing for the inference of cellular functions and properties through a process known as KEGG mapping (Klingström and Plewczynski, 2011).

The Combination of PPI information from protein databases is only acknowledged by SPIKE, leading to the development of their processes among the researched databases. However, ConsusPathDB and current research demonstrate that, in spite of the contrast, data sets from both types can be merged to enhance proteome data while maintaining the standard of useful protein interactions. It must be highlighted, however, that a good amount of data is lost during the merging process, such as semantic links between proteins (Kanehisa and Sato, 2020)

The many ways of characterizing pathway data have hampered data exchange, but it is improving. BioPAX and SBML are the two major data standards for pathway data interchange currently in use. Each has its own set of benefits: SBML supports modeling and simulations, whereas on the other hand, BioPAX creates a quality order with a high adaptability range. As a result, numerous databases accept both formats, with BioPAX dominating in terms of data exchange due to its capacity to communicate a large number of semantic elements (Klingström and Plewczynski, 2011). Another such database is the unifying database, capable of providing single-point entry databases to a number of databases.

4.3.2.5 Unifying database

Apart from DASMI, InterPro and BioMart are similar platforms that provide a single-window entry to a large range of databases and repositories. The InterPro has established its own BioMart service, which allows customers to use the BioMart program to access InterPro. BioMart and DASMI share the goal of providing a singular gateway that allows a majority of the autonomous root datasets. BioMart is more focused on common protein descriptions than just protein interactions. The Reactome, however, provides the BioMart service within its own functions. As a result of using BioMart, one can gain access to InterPro, as well as Protein ANalysis THrough Evolutionary Relationships (PANTHER) Classification System and mapping to IntAct Molecular Interaction Database. The InterPro database organizes amino acid sequences of the proteins into their respective families and finds their conserved regions and functional domains based on this sequence-based annotation. InterPro is a comprehensive database for protein classification that combines signatures identifying comparable sites, domains, and families with supplementary information which includes Gene Ontology (GO) concepts, literature references, descriptions, etc. (Klingström and Plewczynski, 2011) (Blum *et al.*, 2021).

Table 4.2   Links for different databases

Databases	Available at
1. Interaction Databases	
DROID	https://sourceforge.net/projects/droid/
Matrix DB	http://matrixdb.univ-lyon1.fr/
Innate DB	https://www.innatedb.com/
MPI DB	https://eb.mpg.de/
BioGRID	https://thebiogrid.org/

Table 4.2   Continued

Databases	Available at
2.Metamining Databases	
APID	http://cicblade.dep.usal.es:8080/APID/
MiMI	http://www.ncibi.org/mimi.html
UniHI	http://www.unihi.org/
3. Predictive Interaction Databases	
HAPPI	http://discovery.informatics.uab.edu/HAPPI/
STRING	https://string-db.org/
STITCH	http://stitch.embl.de/
ScanSite	https://scansite4.mit.edu/#home
4. Pathway Databases	
KEGG	https://www.genome.jp/kegg/pathway.html
Reactome	https://reactome.org/
5. Unifying Databases	
DASMI	https://www.dasregistry.org/
InterPro	https://www.ebi.ac.uk/interpro/
BioMart	https://m.ensembl.org/info/data/biomart/index.html

4.4 Softwares Available for PPI

The following tables give an abridged version of the various softwares available for studying and analysing PPI networks.

Table 4.3   PPI Prediction tools

	Tool	Principle	Available at	Launch year	Accessibility	References
1	BindML+	Predicts permanent and temporary protein interactions of target protein.	https://kiharalab.org/bindml/plus/	2011	Standalone & Server Based	La and Kihara,
2	PIC	Recognises different types of interactions within proteins or between proteins in a complex.	http://pic.mbu.iisc.ernet.in/	2007	Server Based	Tina et al. 2007
3	MAPPIS	Identifies physio-chemical properties assuming similarity sequential patterns.	http://bioinfo3d.cs.tau.ac.il/MAPPIS/	2007	Server Based	Shulman et al. 2007
4	hotPOINT	Finds hotspots in interacting surfaces of protein via empirical modeling.	http://prism.ccbb.ku.edu.tr/hotpoint/	2010	Server Based	Tuncbag et al., 2011

(Continued)

Table 4.3   Continued

	Tool	Principle	Available at	Launch year	Accessibility	References
5	PSOPIA	Predicts protein interactions based on sequence similarities, statistical propensities and shortest path between homologous proteins.	https://mizuguchilab.org/PSOPIA/	2013	Server Based	Murakmai and Mizuguchi ,2013
6	APID	Integration and analysis of currently known information on proteins.	http://cicblade.dep.usal.es:8080/APID/init.action	2006	Server Based	Prieto and Rivas 2006
7	iFrag	Charting of PPI surfaces.	http://sbi.imim.es/web/index.php/research/servers/iFrag?	2017	Server Based	Garcia *et al.* 2017
8	Interac tome3D	Structural annotation of Protein–protein interaction networks.	https://interactome3d.irbbarcelona.org/index.php	2013	Server Based	Mosca *et al.*, 2013
9	PCRPi-W	Predicts hot spot residues in protein interface.	http://www.bioinsilico.org/PCRPi/	2010	Server Based	Segura *et al.*, 2010
10	PPCheck	Measures the strength of interactions between any two given proteins/chains using standard energy calculations involving non-bonded interactions.	http://caps.ncbs.res.in/ppcheck/	2013	Server Based	Sukhwal and Sowd-hamini 2013
11	SPRINT	Predicts Protein–protein interactions using sequence-based algorithm.	https://github.com/lucian-ilie/SPRINT/	2017	Standalone	Li and Ilie 2017

Table 4.4   Protein binding site prediction tools

S. No.	Tool	Principle	Available at	Year	Accessibility	References
1	META-PPSP	Determines PPI using linear regression method and the raw scores of the three severs as input.	https://pipe.rcc.fsu.edu/meta-ppisp.html	2007	Server based	Qin and Zhou ,2007
2	BindML	Predicts permanent and temporary protein interactions of target protein using the data from the respective protein family MSA.	https://kiharalab.org/bindml/plus/	2011	Standalone & Server Based	La and Kihara 2011.
3	SPPIDER	Predicts interaction sites using an unbound protein 3D structure.	http://sppider.cchmc.org/	2006	Server Based	Porollo and Meller, 2006
4	Inter ProSurf	Analysis of known interactions and identification of interacting surfaces	http://curie.utmb.edu/prosurf.html	2007	Server Based	Negi et al. 2007

Table 4.5 Protein–Protein Prediction tools for model organisms

S.No.	Tool	Principle	Available at	Year	Accessibility	References
1	TRI_tool	Analyses interactions based on sequence information aiming to evade limitations imposed by dispersed auxiliary information.	https://www.vin.bg.ac.rs/180/tools/tfpred.php	2016	Server based	Perovic et al., 2016
2	HIVsemi	Predicts PPIs from not only labelled, but also partially labeled reference sets.	http://www.cs.cmu.edu/~qyj/HIVsemi	2010	Standalone	Qi *et al.* 2010
3	ChiPPI	Predicts location of fusion proteins where they are possible to drop binding to partner proteins of the parent proteins.	http://chippi.md.biu.ac.il/	2017	Server based	Frenkel et al., 2017
4	HIPPIE	Includes scored and functionally annotated human PPIs.	http://cbdm-01.zdv.uni-mainz.de/~mschaefer/hippie/	2012	Server based	Schaefer et al., 2012
5	InteroPORC	Predicts Protein–protein interaction networks for lots of species using orthology and known interactions based on the interolog concept.	http://biodev.extra.cea.fr/interoporc/	2008	Server based & Standalone	Michaut et al., 2008

Table 4.6 Structure Based PPI interaction tool

S.No.	Tool	Principle	Available at	Year	Accessibilty	References
1	InterPreTS	This programme will utilise BLAST to locate homologues of known structure for all pairs of protein sequences (i.e. templates that can model each pair of sequences based on homology) and then assess their suitability for modelling interactions given a collection of protein sequences (in fasta format).	http://www.russelllab.org/cgi-bin/tools/interprets.pl	2002	Server based& Standalone	Aloy & Russell, 2002
2	iLoops	Determines whether or not any two proteins interact using structural properties (protein loops as classified in ArchDB and/or domains as classified in SCOP).	http://aleph.upf.edu/iLoopsServer/	2013	Server based	Joan et al., 2013
3	Struct2Net	Structure-based computational predictions of PPIs.	http://cb.csail.mit.edu/cb/struct2net/webserver/	2010	Server based	R. Singh *et al.* 2010

Table 4.7   PPI Tools for network creation

S.No.	Tool	Principle	Available at	Year	Accessibilty	References
1	**PPI spider**	PPI spider evaluates gene/protein lists using the Global Network statistical framework, with a global Protein–protein interaction network from the IntAct database as a reference.	http://www.bioprofiling.de/PPI_spider.html	2009	Server based	Antonov *et al.* 2009
2	**Path2PPI**	Predict Protein–protein interaction (PPI) networks in target species where only a view of PPIs is available.	http://bioconductor.org/packages/release/bioc/html/Path2PPI.html	2021	Standalone	Philipp,2021

Table 4.8   Other PPI tools

S.No.	Tool	Principle	Available at	Year	Accessibilty	References
1	**FunFOLD2**	The FunFOLD2 server includes the following features: - The list consists of possible binding ligands. Models of potential Protein–ligand interactions in three dimensions	https://www.reading.ac.uk/bioinf/FunFOLD/	2011	Standalone	Roche *et al.* 2011
2	**LIGPLOT**	For a given PDB file, creates schematic representations of Protein–ligand interactions	https://www.ebi.ac.uk/thornton-srv/software/LIGPLOT/	1996	Standalone	Wallace *et al.* 1996
3	**PepSite**	prediction server of small molecules that inhibit Protein–Protein Interactions (PPIs)	http://pepsite2.russelllab.org/	2009	Server based	Petsalaki et al., 2009
4	**PPIMpred**	PPIMpred, or Prediction of Protein–Protein Interaction Modulators, is a webserver that predicts tiny compounds that inhibit Protein–Protein Interactions (PPIs). PIMpred could be used in conjunction with high-throughput docking investigations using tiny molecules. It uses a support vector machine (SVM)-based technique and a similarity search algorithm to quickly and accurately identify ligands on target proteins.	http://bicresources.jcbose.ac.in/ssaha4/PPIMpred/	2017	Server based	Jana et al., 2017

4.5 Conclusion

Computational analysis in biology has always been very important in planning and forming informed hypotheses about biological conditions based on experimental data. These tools have thus helped the researchers to visualize interactions and participation of proteins and other biomolecules without any invasive or experimental biological samples. These tools help the experimental biologist to plan and validate these hypotheses with minimal biological samples. However, these claims based on computational analysis are always subject to experimental validation. These tools also contribute to the multi-omics approaches to find out key solutions to biological problems.

References

Jones, S., and Thornton, J. M. (1996). Principles of Protein–protein interactions. Proceedings of the National Academy of Sciences, 93(1), 13-20.

Rao, V. S., Srinivas, K., Sujini, G. N., and Kumar, G. N. S. (2014). Protein–Protein Interaction Detection: Methods and Analysis. International Journal of Proteomics, 2014, 1–12. doi:10.1155/2014/147648

Keskin, O., Tuncbag, N., and Gursoy, A. (2016). Predicting Protein–protein interactions from the molecular to the proteome level. Chemical reviews, 116(8), 4884-4909.

Klingstrom, T., and Plewczynski, D. (2010). Protein–protein interaction and pathway databases, a graphical review. Briefings in Bioinformatics, 12(6), 702–713. doi:10.1093/bib/bbq064

Ciriello, G., and Guerra, C. (2008). A review on models and algorithms for motif discovery in Protein–protein interaction networks. Briefings in Functional Genomics and Proteomics, 7(2), 147–156. doi:10.1093/bfgp/eln015

Shatnawi M. (2015). Review of Recent Protein–Protein Interaction Techniques. Emerging Trends in Computational Biology, Bioinformatics, and Systems Biology, 99–121. doi:10.1016/b978-0-12-802508-6.00006-5

Lee, S.-A., Chan, C., Tsai, C.-H., Lai, J.-M., Wang, F.-S., Kao, C.-Y., and Huang, C.-Y. F. (2008). Ortholog-based Protein–protein interaction prediction and its application to inter-species interactions. BMC Bioinformatics, 9(Suppl 12), S11. doi:10.1186/1471-2105-9-s12-s11

Hillig, R. C., Sautier, B., Schroeder, J., Moosmayer, D., Hilpmann, A., Stegmann, C. M., Werbeck, N. D., Briem, H., Boemer, U., Weiske, J. and

Badock, V., 2019. Discovery of potent SOS1 inhibitors that block RAS activation *via* disruption of the RAS–SOS1 interaction. Proceedings of the National Academy of Sciences, 116(7), pp.2551-2560.

Xu, S., Aguilar, A., Huang, L., Xu, T., Zheng, K., McEachern, D., Przybranowski, S., Foster, C., Zawacki, K., Liu, Z. and Chinnaswamy, K., 2020. Discovery of M-808 as a highly potent, covalent, small-molecule inhibitor of the Menin–MLL interaction with strong *In Vivo* antitumor activity. Journal of medicinal chemistry, 63(9), pp.4997-5010.

Zhang, G., Andersen, J., and Gerona-Navarro, G. (2018). Peptidomimetics targeting Protein–protein interactions for therapeutic development. Protein and peptide letters, 25(12), 1076-1089.

Jazayeri, A., and Yang, C. C. (2020). Motif discovery algorithms in static and temporal networks: A survey. Journal of Complex Networks, 8(4). doi:10.1093/comnet/cnaa031

Doncheva, N. T., Morris, J. H., Gorodkin, J. and Jensen, L. J., 2018. Cytoscape StringApp: network analysis and visualization of proteomics data. Journal of proteome research, 18(2), pp.623-632.

Oughtred, R., Rust, J., Chang, C., Breitkreutz, B. J., Stark, C., Willems, A., and Tyers, M. (2021). The BioGRID database: A comprehensive biomedical resource of curated protein, genetic, and chemical interactions. Protein Science, 30(1), 187-200.

Szklarczyk, D., Gable, A. L., Nastou, K. C., Lyon, D., Kirsch, R., Pyysalo, S., and von Mering, C. (2021). The STRING database in 2021: customizable Protein–protein networks, and functional characterization of user-uploaded gene/measurement sets. Nucleic acids research, 49(D1), D605-D612.

Kanehisa, M., and Sato, Y. (2020). KEGG Mapper for inferring cellular functions from protein sequences. Protein Science, 29(1), 28-35.

Blum, M., Chang, H. Y., Chuguransky, S., Grego, T., Kandasaamy, S., Mitchell, A., and Finn, R. D. (2021). The InterPro protein families and domains database: 20 years on. Nucleic acids research, 49(D1), D344-D354.

Shen, C. H., 2019. Quantification and Analysis of Proteins in Diagnostic Molecular Biology, Academic Press, pp.187-214.

Guo, Y., Yu, L., Wen, Z., and Li, M. (2008). Using support vector machine combined with auto covariance to predict Protein–protein interactions from protein sequences. Nucleic Acids Research, 36(9), 3025–3030. doi:10.1093/nar/gkn159

Zaki, N., Lazarova-Molnar, S., El-Hajj, W., and Campbell, P. (2009). Protein–protein interaction based on pairwise similarity. BMC Bioinformatics, 10(1). doi:10.1186/1471-2105-10-150

Musheer, R.A., Verma, C.K. and Srivastava, N., 2019. Novel machine learning approach for classification of high-dimensional microarray data. Soft Computing, 23(24), pp.13409-13421.

Valente, G. T., Acencio, M. L., Martins, C., and Lemke, N. (2013). The development of a universal *in silico* predictor of Protein–protein interactions. PloS one, 8(5), e65587.

Wojcik, J., and Schächter, V. (2001). Protein–protein interaction map inference using interacting domain profile pairs. Bioinformatics, 17(suppl_1), S296-S305.

Pazos, F., and Valencia, A. (2001). Similarity of phylogenetic trees as indicator of Protein–protein interaction. Protein engineering, 14(9), 609-614.

Yu, C.-Y., Chou, L.-C., and Chang, D. (2010). Predicting Protein–protein interactions in unbalanced data using the primary structure of proteins. BMC Bioinformatics, 11(1), 167. doi:10.1186/1471-2105-11-167

Valente, G. T., Acencio, M. L., Martins, C., and Lemke, N. (2013). The Development of a Universal *In Silico* Predictor of Protein–Protein Interactions. PLoS ONE, 8(5), e65587. doi:10.1371/journal.pone.0065587

Kern, C., Gonzalez, A. J., Li Liao, and Vijay-Shanker, K. (2013). Predicting Interacting Residues Using Long-Distance Information and Novel Decoding in Hidden Markov Models. IEEE Transactions on NanoBioscience, 123, 158–164. doi:10.1109/tnb.2013.2263810

Yoon, B.-J. (2009). Hidden Markov Models and their Applications in Biological Sequence Analysis. Current Genomics, 10(6), 402–415. doi:10.2174/138920209789177575

Pitre, S., Dehne, F., Chan, A., Cheetham, J., Duong, A., Emili, A., and Golshani, A. (2006). PIPE: a Protein–protein interaction prediction engine based on the re-occurring short polypeptide sequences between known interacting protein pairs. BMC bioinformatics,7(1), 1-15.

Pitre, S., North, C., Alamgir, M., Jessulat, M., Chan, A., Luo, X., and Golshani, A. (2008). Global investigation of Protein–protein interactions in yeast Saccharomyces cerevisiae using re-occurring short polypeptide sequences. Nucleic acids research, 36(13), 4286-4294.

Hsin Liu, C., Li, K. C., and Yuan, S. (2013). Human Protein–protein interaction prediction by a novel sequence-based co-evolution method: co-evolutionary divergence. Bioinformatics, 29(1), 92-98.

Chen, X.-W., and Liu, M. (2005). Prediction of Protein–protein interactions using random decision forest framework. Bioinformatics, 21(24), 4394–4400. doi:10.1093/bioinformatics/bti721

Singh, R., Xu, J., and Berger, B. (2005). Struct2Net: Integrating structure into Protein–protein interaction prediction. biocomputing 2006. doi:10.1142/9789812701626_0037

Pazos, F., and Valencia, A. (2002). *In silico* two-hybrid system for the selection of physically interacting protein pairs. Proteins: Structure, Function, and Genetics, 47(2), 219–227. doi:10.1002/prot.10074

Ozbabacan, S. E. A., Keskin, O., Nussinov, R., and Gursoy, A. (2012). Enriching the human apoptosis pathway by predicting the structures of Protein–protein complexes. Journal of structural biology, 179(3), 338-346.

Tuncbag, N., Kar, G., Gursoy, A., Keskin, O., and Nussinov, R. (2009). Towards inferring time dimensionality in Protein–protein interaction networks by integrating structures: the p53 example. Molecular Biosystems, 5(12), 1770-1778.

Zhang, Q. C., Petrey, D., Deng, L., Qiang, L., Shi, Y., Thu, C. A., and Honig, B. (2012). Structure-based prediction of Protein–protein interactions on a genome-wide scale. Nature, 490(7421), 556-560.

Poluri, K. M., Gulati, K., and Sarkar, S. (2021). Prediction, Analysis, Visualization, and Storage of Protein–Protein Interactions Using Computational Approaches. In Protein–Protein Interactions (pp. 265-346). Springer, Singapore.

Han, D. S., Kim, H. S., Jang, W. H., Lee, S. D., and Suh, J. K. (2004). PreSPI: a domain combination based prediction system for Protein–protein interaction. Nucleic acids research, 32(21), 6312-6320.

Ohue, M., Matsuzaki, Y., Uchikoga, N., Ishida, T., and Akiyama, Y. (2014). MEGADOCK: an all-to-all Protein–protein interaction prediction system using tertiary structure data. Protein and peptide letters, 21(8), 766-778.

Ohue, M., Shimoda, T., Suzuki, S., Matsuzaki, Y., Ishida, T., and Akiyama, Y. (2014). MEGADOCK 4.0: an ultra–high-performance Protein–protein docking software for heterogeneous supercomputers. Bioinformatics, 30(22), 3281-3283.

Ishida, T., and Kinoshita, K. (2008). Prediction of disordered regions in proteins based on the meta approach. Bioinformatics, 24(11), 1344-1348.

La, D., and Kihara, D. (2011). A novel method for Protein–protein interaction site prediction using phylogenetic substitution models.

Proteins: Structure, Function, and Bioinformatics, 80(1), 126–141. doi:10.1002/prot.23169

K. G. Tina, R. Bhadra and N. Srinivasan, PIC: Protein Interactions Calculator, Nucleic Acids Research, 2007, Vol. 35, Web Server issue W473–W476.

Shulman-Peleg A, Shatsky M, Nussinov R. and Wolfson H.J. Spatial chemical conservation of hot spot interactions in Protein–protein complexes. BMC Biol. 2007 Oct 9;5(1):43

Tuncbag, N., Keskin, O., and Gursoy, A. (2010). HotPoint: hot spot prediction server for protein interfaces. Nucleic Acids Research, 38(Web Server), W402–W406. doi:10.1093/nar/gkq323

Prieto, C., and De Las Rivas, J. (2006). APID: agile protein interaction DataAnalyzer. Nucleic acids research, 34(suppl_2), W298-W302.

Garcia-Garcia, J., Valls-Comamala, V., Guney, E., Andreu, D., Muñoz, F. J., Fernandez-Fuentes, N., and Oliva, B. (2017). iFrag: a Protein–protein interface prediction server based on sequence fragments. Journal of molecular biology, 429(3), 382-389.

Mosca R, Céol A, Aloy P, Interactome3D: adding structural details to protein networks, Nature Methods (2013) 10(1):47-53, doi:10.1038/nmeth.2289

Segura Mora, J., Assi, S. A., and Fernandez-Fuentes, N. (2010). Presaging critical residues in protein interfaces-web server (PCRPi-W): a web server to chart hot spots in protein interfaces. PloS one, 5(8), e12352.

Sukhwal, A.; Sowdhamini, R. Oligomerisation status and evolutionary conservation of interfaces of protein structural domain superfamilies. Mol. Biosyst. 2013, 9, 1652-1661

Li, Y., and Ilie, L. (2017). SPRINT: ultrafast Protein–protein interaction prediction of the entire human interactome. BMC Bioinformatics, 18(1). doi:10.1186/s12859-017-1871-x

La, D., and Kihara, D. (2011). A novel method for Protein–protein interaction site prediction using phylogenetic substitution models. Proteins: Structure, Function, and Bioinformatics, 80(1), 126–141. doi:10.1002/prot.23169

Qi, Y., Tastan, O., Carbonell, J. G., Klein-Seetharaman, J., and Weston, J. (2010). Semi-supervised multi-task learning for predicting interactions between HIV-1 and human proteins. Bioinformatics, 26(18), i645–i652. doi:10.1093/bioinformatics/btq394

Singh, R., Park, D., Xu, J., Hosur, R., and Berger, B. (2010). "Struct2Net: a Web-Service to Predict Protein–Protein Interactions Using a Structure-based Approach." Nucleic Acids Research, doi:10.1093/nar/gkq481.

Antonov, A. V., Dietmann, S., Rodchenkov, I., Mewes, H. W. PPI spider: A tool for the interpretation of proteomics data in the context of protein

protein interaction networks. PROTEOMICS. Volume 9, Issue 10, 10 May 2009.

Roche, D. B., Tetchner, S. J. and McGuffin, L. J. (2011) FunFOLD: an improved automated method for the prediction of ligand binding residues using 3D models of proteins. BMC Bioinformatics, 12, 160.

Wallace, A. C., Laskowski, R. A., Thornton, J. M. (1996). LIGPLOT: a program to generate schematic diagrams of Protein–ligand interactions. Protein Eng., 8, 127-134.

Hosur, R., Xu, J., Bienkowska, J., & Berger, B. (2011). iWRAP: an interface threading approach with application to prediction of cancer-related protein–protein interactions. Journal of molecular biology, 405(5), 1295-1310.

5

Optimization of COVID-19 Risk Factors Using Fuzzy Logic Inference System

Jyoti Gupta*, Samidha Saxena, Namrata Kaushal, and Parimeeta Chanchani

Department of Engineering Science & Humanities, Indore Institute of Science & Technology, Indore, MP, India
E-mail: jyoti.gupta@indoreinstitute.com
*Corresponding Author

Abstract

According to the WHO, the topmost priority during the COVID-19 pandemic is to recognize the risk factors for the severity of this disease. Because of this, we conducted a series of calculations based on several symptoms of CORONA infections. The study aimed to estimate the rigorousness and identify the risk factors of COVID-19 infection and calculated the risk % by taking a wide range of symptoms like Body Temperature, Cough, Cold, Breathing problems, and Loss of senses of Smell & Taste. We have used MATLAB to simulate a model based on Mamdani fuzzy inference system to help those who can identify their symptoms. In the proposed model Mean of Maxima kind of De-Fuzzifier is applied. Additionally, we also conducted a comparability analysis of risk factors across 5 - 6 studies. The study concludes that if a patient's body temperature is 38.4 °C, suffering from cough (6), cold (8) and breathing rate in pulse oxy-meter is 95, loss of sense of smell is 17% then the risk of his being infected by coronavirus is 50%. Based on the results obtained, we have also proposed a set of rules for further prevention and mitigation of pandemics.

Our findings will help in developing targeted prevention and control strategies to combat this worldwide pandemic. In the future also the outcomes

101

are very beneficial when Artificial Neural networks, Machine Learning will be used to train the model and provide more accurate results. The results have also motivated the authors that the inter-disciplinary approach toward such collaborative research works would lead to finding more effective solutions to such serious problems. In this era when there is the threat of biological warfare across the globe, such studies have also opened new avenues for strategic & timely mitigation of biological agents over large sample sizes.

Keyword: COVID-19, Risk Factors, MATLAB, Mamdani fuzzy inference system, De-Fuzzifier.

5.1 Introduction

Today, the COVID-19 pandemic is one of those global challenges that go beyond territorial, political, conceptual, religious, cultural, and certainly speculative boundaries. Almost all the countries and the regions have reported confirmed cases of the novel coronavirus disease COVID-19, which is already considered a pandemic by the World Health Organisation on April 7, 2020. As these global health emergency checks, the pliability of health systems around the globe, health care, and public health practitioners are mandatorily needed to have excellent evidence to identify its most substantial risks and prioritize resources where they are most desirable. One of the most important demands to address the presently unfolding pandemic is "what are the threat factors for severe illness or death?" [1, 2]. Organized data collections, reviews, and peered analysis paired with the standardized method to evaluate the quality of evidence are deemed to provide the best evidence [3]. Contact tracing is found to be a major cause of the imports of some rare and emerging infectious diseases. The main objective of contact tracing is to identify potentially infected individuals before the onset of severe symptoms, and also to prevent the onward transmission from the secondary cases. Contact tracing has decisively contributed to the control of numerous contagious diseases worldwide including severe acute respiratory syndrome (SARS), Ebola virus disease, and Middle East respiratory syndrome (MERS) [4–7]. Previous studies using mathematical modeling demonstrated that contact tracing and quarantine play important roles in controlling the spreading of COVID-19 [8, 9, 25] Researchers have analyzed the data of COVID-19 patients and found some risk factors for mortality, such as older age, pre-existing cardiovascular or cerebrovascular diseases, low levels of $CD3^+CD8^+$ T-cells, and high levels of cardiac troponin I, a higher Sequential Organ Failure Assessment score and d-dimer [11, 12]. Unfortunately, limited study has paid attention to the risk factors related to COVID-19 infection. Some recent studies identified the exposure to index cases with severe symptoms as a risk factor [13] [17–20] Still, their limited sample size, especially the limited cases, may restrict their capability to perform detailed analysis and reduce the power to detect significant risk factors. Also, findings within a small region or selected sample may restrict its ability to conception. An alternative method utilizing waste water-based epidemiology (WBE), has also been developed as an effective approach to forecasting the implicit spread of the infection and the risk proportion by testing for infectious agents in wastewater [14, 15].

Since, scientific publications revealing the pandemic study outcomes are being produced rapidly, including risk factor studies, calculating and sharing our findings is of paramount significance to support an efficient and speedy response. An early review of risk factor studies could also provide some perception of the undesirable heterogeneity of definitions and reporting that might affect later evidence estimations. Therefore, we conducted some analysis using MATLAB and the Fuzzy Inference model based on various symptoms to predict the risk % for Covid-19 disease severity. To the best of our knowledge, there are very few studies based on Fuzzy interface modeling are carried out and we hope that this research will proactively help to understand the challenges, develop new hypotheses and define new research areas [26-28].

5.2 Methodology

5.2.1 Proposed Mamdani Fuzzy Control System

In this process, we have carried out an analysis with a fuzzy rule-based model under Mamdani fuzzy inference system. The main idea of this system was to describe process states employing linguistic variables and to use these variables as inputs to control rules. This system predicted the risk percentage whether the patient was suffering from coronavirus by comparing his symptoms with the symptoms of COVID-19 as declared by World Health Organization. The purpose of this model is mainly to make people able to check their symptoms and to find what percentage of risk they might have when being infected by COVID-19.

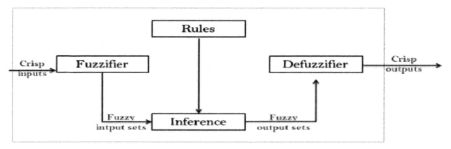

Figure 5.1   Fuzzy Control System

Table 5.1   Parameters and membership values with linguistic ranges

Parameters	Membership values with linguistic ranges		
Body Temperature (BT)	Low	Medium	High
Cough	Yes	No	
Cold	Yes	No	
Breathing Problem (BrP)	Low	Medium	High
Loss of sense of Smell/Taste (LSS)	Low	Medium	High
Risk%	Low	Medium	High

5.2.2 Fuzzy Controller Design

For the implementation of this proposed model MATLAB is used with a fuzzy logic control tool. Following are the steps of the proposed fuzzy rule-based system:

5.2.3 Parameters Identification

We have designed this fuzzy inference system by taking five symptoms of covid-19 as input; Body Temperature (BT), cough, cold, Breathing Problem (BrP), and Loss of Sense of Smell/taste (LSS) with one output risk percentage. Each of these parameters is further classified into three linguistic ranges, which denoted the risk level of the patient being infected by a coronavirus.

5.2.4 Fuzzification

The fuzzification function was introduced for each input variable to express the associated range. In the fuzzy inference system tool, each input is taken in the form of membership function (MF) represented between [0-1] intervals and expressed as μ. In this paper, we have used triangular and trapezoidal membership functions for each input and output variable as shown the Figures 5.2, 5.3, 5.4, 5.5 and 5.6.

(i) Body Temperature (BT)

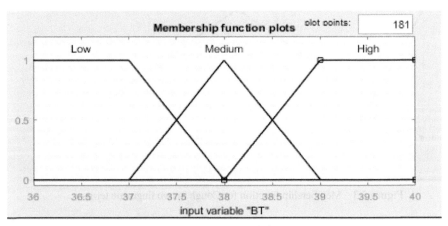

Figure 5.2 Membership function for Body Temperature in three linguistic terms

The membership equation and the value of μ for body temperature of each linguistic term are:

$$\mu_{low}(x) = \begin{cases} 1; & x<37\,°C \\ 38-x; & 37\,°C \leq x \leq 38\,°C \\ 0; & x>38°C \end{cases}$$

$$\mu_{Medium}(x) = \begin{cases} 0; & x<37\,°C \\ x-37; & 37\,°C \leq x \leq 38\,°C \\ 39-x; & 38\,°C \leq x \leq 39\,°C \end{cases}$$

$$\mu_{High}(x) = \begin{cases} 0; & x<38\,°C \\ x-38; & 38\,°C \leq x \leq 39\,°C \\ 1; & x>39\,°C \end{cases}$$

(ii) Cough

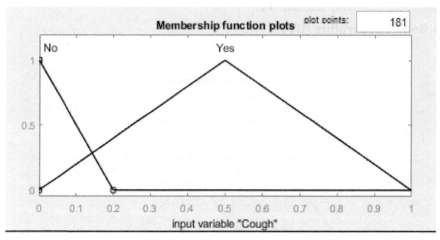

Figure 5.3   Membership function for Cough in two linguistic terms

The membership equation and the value of μ for Cough of each linguistic term are:

$$\mu_{No}(x) = \begin{cases} \frac{0.2-x}{0.2}; & x < 0.2 \\ 0; & x \geq 0.2 \end{cases}$$

$$\mu_{Yes}(x) = \begin{cases} \frac{x}{0.5}; & 0 \leq x \leq 0.5 \\ \frac{1-x}{0.5}; & x > 0.5 \end{cases}$$

(iii) Cold

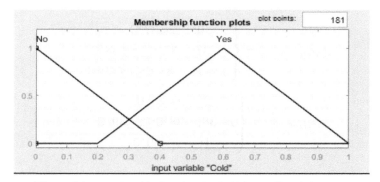

Figure 5.4   Membership function for Cold in two linguistic terms

The membership equation and the value of μ for Cold of each linguistic term are:

$$\mu_{No}(x) = \begin{cases} \frac{0.4-x}{0.4}; & x<0.4 \\ 0; & x\geq0.4 \end{cases}$$

$$\mu_{Yes}(x) = \begin{cases} 0; & x<0.2 \\ \frac{x-0.6}{0.4}; & 0.2\leq x\leq0.6 \\ \frac{1-x}{0.4}; & x>0.6 \end{cases}$$

(iv) Breathing Problem (BrP)

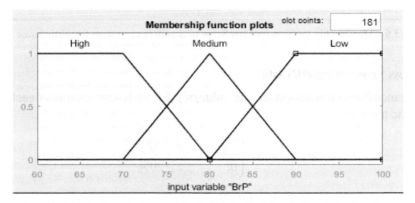

Figure 5.5 Membership function for Breathing Problem in three linguistic terms

The membership equation and the value of μ for body temperature of each linguistic term are:

$$\mu_{High}(x) = \begin{cases} 1; & x<70 \\ \frac{80-x}{10}; & 70\leq x\leq80 \\ 0; & x>80 \end{cases}$$

$$\mu_{Medium}(x) = \begin{cases} 0; & x<70 \\ \frac{x-70}{10}; & 70\leq x\leq80 \\ \frac{90-x}{10}; & 80\leq x\leq90 \end{cases}$$

$$\mu_{Low}(x) = \begin{cases} 0; & x<80 \\ \frac{x-80}{10}; & 80\leq x\leq90 \\ 1; & x>90 \end{cases}$$

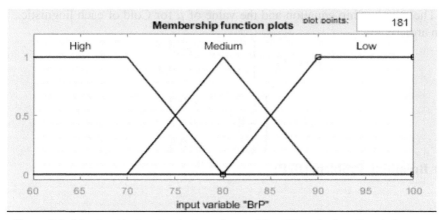

Figure 5.6 Membership function for Loss Sense of Smell/Taste in three linguistic terms

(v) Loss Sense of Smell/Taste

The membership equation and the value of μ for body temperature of each linguistic term are:

$$\mu_{Low}(x) = \begin{cases} \frac{20-x}{20}; & x<20\% \\ 0; & x\geq20\% \end{cases}$$

$$\mu_{Medium}(x) = \begin{cases} \frac{x-10}{10}; & 10\%\leq x\leq20\% \\ \frac{30-x}{10}; & 20\%\leq x\leq30\% \\ 0; & x>30\% \end{cases}$$

$$\mu_{High}(x) = \begin{cases} 0; & x<20\% \\ \frac{x-20}{20}; & 20\%\leq x\leq40\% \\ 1; & x>40\% \end{cases}$$

5.2.5 Fuzzy Inference Rule Base

The fuzzy inference rules connect the input variables with the output variables and are based on the fuzzy state description which is obtained by the definition of the linguistic variables. In this paper, we have used five input variables consisting of body temperature, cough, cold, breathing problem, loss of sense of smell/taste and each variable (BT, BrP, LSS) has been divided into 3 linguistic categories and input variables (cough & cold) have been divided into 2 linguistic categories. According to Allahverdi and Ertosun (2018) [16],

Table 5.2   Fuzzy logic-based rules

Input Variables					Output
Body temperature (BT)	**Cough**	**Cold**	**Breathing Problem (BrP)**	**Loss of sense of smell/taste (LSS)**	**Risk%**
Low	No	No	Low	Low	Low
High	Yes	Yes	Low	Low	Medium
Low	Yes	No	Medium	Medium	Medium
High	Yes	Yes	High	Low	High
Medium	No	Yes	Low	Low	Low
Medium	Yes	Yes	Low	Low	Medium
Low	No	No	Medium	Low	Low
High	No	No	Medium	Medium	Medium
High	Yes	Yes	High	High	High
Medium	Yes	Yes	Low	Medium	Medium
High	Yes	Yes	Low	Medium	High
...	...	...	...	...	...
...	...	...	...	...	...
...	...	...	...	...	...
Medium	Yes	Yes	High	High	High

the numbers of fuzzy rule combinations have been calculated as $3*3*3*2*2 = 108$ rule combinations. We have constructed the If... then rules by using the 'and' operator in the antecedent.

5.2.6 Rule Evaluation By Fuzzy Inference Engine

Based on the linguistic range of input symptoms inference engine properly combines the relevant fuzzy rules and then determines the degree of membership of the input in the rule antecedent. Employing the minimum operator as a model for the "and," we have computed the degree of membership of each relevant rule. The computation of the output risk% of coronavirus is obtained by aggregation of all consequences using the maximum operator. Figure 5.7 shows the structure of the fuzzy inference system.

5.2.7 Defuzzification

In this last step of the proposed system process, the fuzzy controller must select a suitable defuzzification method. The purpose of defuzzification is to convert each conclusion obtained by the inference engine, which is expressed in terms of a fuzzy set, to a crisp value. The membership equation and the

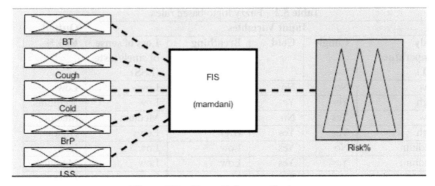

Figure 5.7   Fuzzy Inference System

value of μ for output variable risk% of each linguistic term are:

$$\mu_{Low}(z) = \begin{cases} \frac{30-z}{20}; & x \leq 30\% \\ 0; & x > 30\% \end{cases}$$

$$\mu_{Medium}(z) = \begin{cases} \frac{z-30}{20}; & 30\% \leq x \leq 50\% \\ \frac{70-z}{20}; & 50\% \leq x \leq 70\% \\ 0; & x > 70\% \end{cases}$$

$$\mu_{High}(z) = \begin{cases} 0; & x < 70\% \\ \frac{x-70}{10}; & 70\% \leq x \leq 80\% \\ 1; & x > 80\% \end{cases}$$

In this model, we have used the "Mean of Maxima" method for defuzzification of the risk%, which is the average of all values in the crisp set (Risk%) defined by

$$d_{MM}(Risk\%) = \frac{\sum z}{n}$$

5.3 Results

The result of the proposed fuzzy system in this study is obtained through the input of real parameter values that if the body temperature is 38.4 °C, suffering from cough (6) and cold (8), breath rate in pulse oxy meter is 95, loss of sense of smell 17%, then the patient will have 50% risk of being infected by a coronavirus. In MATLAB this result can be shown through the rule view table.

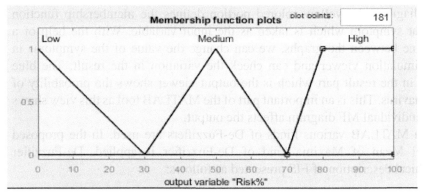

Figure 5.8 Membership function for Risk% in three linguistic terms

Table 5.3 Risk% of the infected patient

	BT = 38.4	Cough = 0.602	Cold = 0.809	BrP = 95.3	LSS = 17.1	Risk% = 50
1						
2						
3						
4						
5						
6						
7						
8						
9						
10						
11						
12						
13						
14						
15						
16						
17						
18						
19						
20						
	36 40	0 1	0 1	60 100	0 50	

The above table shows the description of rules that are well-defined by the possible symptoms of the coronavirus. It is the entire system of inference process defined for the prediction risk of the virus. With the help of the rule viewer, the fuzzy inference system diagram of the model can be viewed. In this diagram, the yellow-colored portion defines the membership function of that symptom which is taken as the input variable. With the help of a red line between the graphs, we can change the value of the symptoms in the simulation viewer and can check the variation in the result. The blue color in the result part which is the output viewer shows the probability of coronavirus. This is an important part of the MATLAB tool as this view shows how individual MF diagram affects the output.

In MATLAB various kinds of De-Fuzzifiers are used. In the proposed model Mean of Maxima kind of De-Fuzzifier is applied. De-Fuzzifier graphical description of FIS presented as follows:

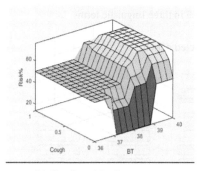

(a) Cough and Body temperature

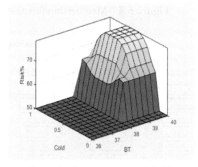

(b) Body Temperature and Cold

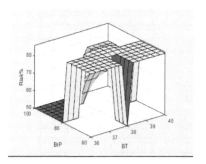

(c)Body temperature and Breathing Problem

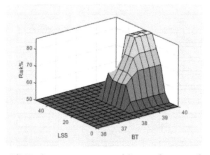

(d) Body temperature and Loss of sense of smell/taste

Figure 5.9　3D form Rule Surface

The above figures show representation in 3D form as a ruled surface performed in MATLAB, which represents a mapping from two input values (symptoms) to the risk%. The color Blue represents simulation weak, Green as satisfied, and yellow as good.

5.4 Discussion

The proposed fuzzy inference system in this paper will not only support the doctors in their prediction but also for every human being to predict the risk% of infection of coronavirus based on their symptoms. As we observe that the spread of covid-19 is now rapidly increasing at the community level in India and many other parts of the world and our corona warriors are fighting at the cost of their own lives to handle the situation. This model can be used by people to self-monitor the risk% of covid-19 infection and to quarantine themselves if they are at low or medium risk. There might be numerous symptoms that can affect any person if they have weak immunity like fever, cough, cold, breathlessness, and difficulty smelling or tasting. A person who has traveled through bus, train, flight, or self-vehicle from a wildly spread place of coronavirus and if anyone is having any of these symptoms then it is suggested by WHO to ensure self-assessment and proper precautions should be taken to avoid any kind of more complications.

Now we deliberate the working process of our proposed model by considering the input of real parameters like body temperature of 38.4 °C, suffering from cough (6) and cold (8), breathing rate in pulse oxy meter is 95, loss of sense of smell 17%. Following are the steps of calculation through the system:

1. Fuzzification of body temperature membership (Figure 5.2) at $x=$ 38.4 °C yields $\mu = 0.6$ for medium body temperature fuzzy sets and $\mu = 0.4$ for high body temperature fuzzy sets. Fuzzification of cough (Figure 5.3) and cold (Figure 5.4) at $x= 0.6$ yields $\mu = 0.8$ and at $x= 0.8$ yields $\mu = 0.5$ respectively.

2. While the fuzzification of breathing problem (Figure 5.5) at $x= 95$ yields $\mu = 1$ for low BrP fuzzy set and loss of sense of smell/taste (Figure 5.6) at $x= 17\%$ yields $\mu = 0.15$ for low LSS fuzzy set and $\mu = 0.7$ for medium LSS fuzzy set.

3. Now the system will define the rules on basis of symptoms as follows:

   (i) If the patient's body temperature is medium and cough (yes) and cold (yes) and breathing problem is low and loss of sense of smell/taste is low, then the risk is medium.

   (ii) If the patient's body temperature is medium and cough (yes) and cold (yes) and breathing problem is low and loss of sense of smell/taste is medium, then the risk is medium.

(iii) If the patient's body temperature is high and cough (yes) and cold (yes) and breathing problem is low and loss of sense of smell/taste is low, then the risk is medium.

4. If the patient's body temperature is high and cough (yes) and cold (yes) and breathing problem is low and loss of sense of smell/taste medium, then the risk is high.

(i) Based on the rule (i), the fuzzification values are "If patient's body temperature is medium ($\mu = 0.6$) and cough (yes) ($\mu = 0.8$) and cold (yes) ($\mu = 0.5$) and breathing problem low ($\mu = 1$) and loss sense of smell/taste low $\mu = 0.15$, then the risk is medium", and the medium risk is the minimum value of μ equal to 0.15. Similarly from the rule (ii), (iii), (iv), the membership value μ for the risk is equal to 0.5, 0.15, and 0.4.

(ii) The next step is defuzzification, since we have used the "Mean of Maxima" defuzzifier, the maximum risk membership value μ is equal to max$\{0.15, 0.5, 0.15, 0.4\}= 0.5$, which is corresponding to rule (ii) shows the risk is medium.

(iii) To find out the final defuzzification value, the system takes the mean of μ_{Medium} for risk.

Since

$$\mu_{Medium}\,(z)=\frac{z-30}{20} \Rightarrow\ 0.5 =\frac{z-30}{20}\Rightarrow z= 40\ \text{and}$$

$$\mu_{Medium}\,(z)=\frac{70-z}{20} \Rightarrow\ 0.5 =\frac{70-z}{20}\Rightarrow z= 60$$

Thus

$$\text{Mean of risk \%}=\frac{40+60}{2}= 50\%$$

The risk information would be helpful to make a decision and take essential preventive measures. Precautions suggested by WHO are also included in this proposed model, as when an infected person does his self-assessment and if there is any possibility of coronavirus then the model suggests what kind of precautions should be taken care of.

Following are the precautions that should be taken:

1. Wash your hands frequently
2. Maintain social distancing
3. Use a mask if suffering from cough or cold
4. Do not touch the eyes, nose, or mouth

5. Hygiene should be taken care of
6. If having a high fever with cough and cold then go for a check-up.

5.5 Conclusion and Future Work

It is clear that the threat of COVID-19 outbreak is not confined to any single country or region and also its symptoms are dynamic and can have affected by various parameters, it is necessary to establish some guidelines for the responses, control, and prevention of this novel infectious disease. Based on risk factors observations, it can be concluded that the Novel Coronavirus remains on the human body for 14 days and during the early stages of infection there are no symptoms but if the patient has any medical history, chronic problem, or might have traveled to the infected country or might come in contact with an infected person then the person should have to do this assessment daily up to 14 days. This model can only help those who can identify their symptoms, but if someone is ignorant about the virus threat and has been exposed somewhere to other non-infected people then it is highly believed that the other persons might be infected by that person.

In the present work, we have used MATLAB to simulate our model. In the future Artificial Neural networks and Machine Learning can also be used to get more accurate results and more relevant parameters for a better understanding of the disease. Secondly for other deadly diseases also where linguistic variables are well defined we can develop such models for timely prevention. For those who don't understand the symptoms of diseases, some sensor-based devices can be developed using such models in the future. This will not only help us in segregating the population-based on criticality and unnecessary rush in hospitals can be avoided.

References

[1] World Health Organization. Coronavirus disease 2019 (COVID-19) Situation Report – 78, April 2020.
[2] Lipsitch M, Swerdlow DL, Finelli L. Defining the epidemiology of Covid-19: studies needed, N Engl J Med. 2020; 382:1194-1996. https://Doi.10.1056/NEJMp2002125.
[3] Murad MH, Asi N, Alsawas M, Alahdab F. New evidence pyramid. Evid Based Med. 2016; 21(4):125-127. https://Doi.10.1136/ebmed-2016-1 10401.

[4] Rod J. E., Trespalacios O. O., and Cortes-Ramirez J., A brief review of the risk factors for covid-19 severity, Rev Saude Publica 2020; 54: 60. ~https://Doi.10.11606/s1518-8787.2020054002481.

[5] Glasser JW, Hupert N, McCauley M., et al., Modeling, and Public health emergency responses: Lessons from SARS Epidemics 2011; 3(1):32–37. https://doi:10.1016/j.epidem.2011.01.001.

[6] Pang X., Zhu Z., Xu F., et al. Evaluation of control measures implemented in the severe acute respiratory syndrome outbreak in Beijing, 2003, JAMA, 2003; 290(24):3215–3221. https://doi:10.1001/jama.290.24.3215.

[7] Hellewell J., Abbott S., Gemma A., et al. Feasibility of controlling COVID-19 outbreaks by isolation of cases and contacts. Lancet Global Health. 2020; 8(4):e488–e496. https://doi:10.1016/S2214-109X(4)30074-7.

[8] Keeling MJ., Hollingsworth TD., Read JM., The efficacy of contact tracing for the containment of the 2019 novel coronavirus (COVID-19). medRxiv. 2020. https://doi:10.1101/2020.02.14.20023036.

[9] Du RH., Liang L R., Yang CQ., et al. Predictors of mortality for patients with COVID-19 pneumonia caused by SARS-CoV-2: a prospective cohort study. Eur Respir J. 2020;55(5). https://doi:10.1183/13993003.00524-2020.

[10] ZhouF, YuT, DuR, et al. Clinical course and risk factors for mortality of adult inpatients with COVID-19 in Wuhan, China: a retrospective cohort study. Lancet. 2020;395(10229):1054–1062. https://doi:10.1016/S0140-6736(20)30566-3.

[11] Mao K., Zhang H. and Yang Z. Can a Paper-Based Device Trace COVID-19 Sources with Wastewater-Based Epidemiology? Environ. Sci. Technol., 2020; 54(7):3733–3735. http://doi:10.1021/acs.est.0c01174.

[12] Allahverdi N., Ertosun S., Application of fuzzy logic for risk determination of type 2 diabetes disease, 6th International Conference on Control and Optimization with Industrial Applications, July 2018.

[13] Painuli D., Mishra D., Bhardwaj S., Agrawal M., Fuzzy Rule-Based System to predict COVID19 - A Deadly Virus, Inter. J. Management & Humanities, 2020; 4(8);78-82. http://doi:10.35940/ijmh.H0781.044820.

[14] Katigari, M. R., Ayatollahi, H., Malek, M., and Haghighi, M. K. "Fuzzy expert system for diagnosing diabetic neuropathy", 2015. World J Diabetes, 8(2):80-88. http://doi:10.4239/wjd.v8.i2.80.

[15] Thukral. S., Medical Applications on Fuzzy Logic Inference System: A Review, Int. J. Advanced Networking and Applications, 2019; 10(4):3944-3950.

[16] Adwibowo A., Fuzzy logic assisted COVID 19 safety assessment of dental care,medRxiv, 2020. https://doi.org/10.1101/2020.06.18.201 34841.

[17] Spagnuolo G., De Vito D., Rengo S., Tatullo M., COVID-19 Outbreak: An Overview on Dentistry, Int. J. Environ. Res. Public Health, 2020; 17: 2094.

[18] Makrariya, Akshara, and Kamal Raj Pardasani. "Numerical simulation of thermal changes in tissues of woman's breast during the menstrual cycle in different stages of its development." International Journal of Simulation and Process Modelling 14.4 (2019): 348-359.

[19] Dhiman N and Sharma M. K., "Mediative Sugeno's-TSK fuzzy logic-based screening analysis to the diagnosis of heart disease", Applied Mathematics, 2019; 10:448-467. http://doi:10.4236/am.2019.106032.

[20] Nosyk B., Armstrong W. S., Rio C. D., Contact Tracing for COVID-19: An Opportunity to Reduce Health Disparities and End the Human Immunodeficiency Virus/AIDS Epidemic in the United States, Clin Infect Dis. 2020; 71(16): 2259–2261. https://doi:10.1093/cid/ciaa501.

[21] Kucharski A. J., Klepac P. Conlan A. JK., et al., Effectiveness of isolation, testing, contact tracing and physical distancing on reducing transmission of SARS-CoV-2 in different settings: a mathematical modeling study, 2020; 20(10):1151-1160. https://doi.org/10.1016/S1473-30 99(20)30457-6.

[22] Aziz R., Verma C. K., Srivastava N. Artificial neural network classification of high dimensional data with novel optimization approach of dimension reduction. Annals of Data Science. 2018 Dec;5(4):615-35.

[23] Aziz R., Verma C., Srivastava N. A weighted-SNR feature selection from independent component subspace for nb classification of microarray data. Int J Adv Biotec Res. 2015 Jan 1;6:245-55.

[24] Aziz R., Srivastava N., Verma C. K. T-independent component analysis for SVM classification of DNA-micorarray data. International Journal of Bioinformatics Research, ISSN. 2015:0975-3087.

6

Dynamical Analysis of the Fractional-Order Mathematical Model of Hashimoto's Thyroiditis

Neelam Singha

Department of Mathematics, School of Technology, Pandit Deendayal Energy University, Gandhinagar, Gujarat-382426, India

Abstract

The present work addresses a fractional-order mathematical model handling an auto-immune condition against the thyroid follicle cells, Hashimoto's thyroiditis (HT). Under this condition, the thyroid-stimulating hormone (TSH) changes more rapidly than the free thyroxine (FT4), resulting in the thyroid follicle cells getting destroyed by the interrupted working process of the hypothalamus- pituitary-thyroid (HPT) axis. We address the modernization of an existing 4d-model of HT by incorporating the fractional-order operators. The proposed fractional-order model comprises four time-dependent variables namely, TSH, FT4, anti-thyroid antibodies (Ab), and size of the thyroid gland (T). We discuss the local stability conditions for the recommended problem of fractional-order, followed by constructing a numerical solution scheme. We implement the Adomian decomposition method (ADM) on the extended fractional model of HT to retrieve the approximate solutions. In addition, we plot the time-dependent state variables involved to interpret the numerical results effectively.

Keywords: Fractional derivative, Hashimoto's thyroiditis, Mathematical modeling, Adomian decomposition method 2010 MSC: 26A33, 35R11, 44A55, 47G20.

119

6.1 Introduction

Analyzing bio-mathematical models of harmful diseases, encompassing complex structural processes, is a progressive research area that is challenging and compelling to present day researchers. For almost a century, it has undoubtedly fascinated the experts to understand the dynamics of a disease quantitatively by exploring computational mathematical models. And, the mathematical characterization of any biological disease is described by multiscale factors, which are required to be discussed significantly based on the available information about the system. One of the elementary benefits of these models is the estimation of essential parameters bearing the competency to furnish important aspects not directly accounted for in the system. We expect that the crucial details obtained through these models will improve the present understanding of the concerned disease and its treatment.

Over the past years, the process of any biological system has been modeled mathematically by ordinary differential equations to interpret the system dynamics efficiently. For instance, mathematical models of tumor-immune interactions [1, 2], dengue and malaria transmission [3, 4], epidemiology of infectious diseases [5, 6], etc. One may also notice that most the biological systems acknowledge natural history, and an organized study of related memory will surely help to perceive a comparatively better understanding of important aspects of any disease.

Recently, non-integer order derivatives are adopted frequently to depict the memory preserving behavior, which remains neglected by classical derivatives. The notion of fractional-order derivatives existed in the literature for over 300 years but initially encountered differences owing to the inadequate geometrical and physical assimilation. Later, many researchers (see [7, 8, 9] and [10]) presented the appropriate descriptions for fractional-order operators. Readers interested in the systematic historical development of fractional calculus may refer to [11, 12]. Subsequently, the area of fractional calculus acquired its due importance because of its applicability in diversified branches of engineering and medical sciences (see [13, 17] and references therein). For a detailed report on various existing applications, we refer the reader to [18, 19].

In contrast to integer-order bio-mathematical models, fractional-order models are now being used to bring in the memory effects due to their non-local behavior. A part of the exploration and experimentation work rendered in this area covers computational modeling in complex biological systems

[20, 21, 22], fractional-order model of cancer [23, 24] and HIV [25], practical administration of fractional optimal control problems [26, 27, 28, 29], population growth model [17], etc.

Through the presented work, we plan to describe an integer-order model of Hashimoto's thyroiditis (HT) (see [30, 31],) with the assistance of fractional-order derivatives. HT refers to an autoimmune disease conferred by the thyroid gland in which the immune system gradually attacks the thyroid follicle cells. As a result, the thyroid becomes incapable of supplying the adequate amount of hormones required for the normal axis operation. One may refer to [32] for a thorough explanation of the physiology of the HT and the existing integer-order mathematical models.

We start with extending the integer-order model of HT into a memory pre- serving model with a finite set of arbitrary order differential equations. The primary contribution of the stated work is analyzing the dynamics of the proposed model to have deep insights into the system dynamical process appearing in the fractional-order model of HT. Next, we intend to execute a stability analysis to identify the conditions under which the modified model becomes stable. And, the resulting model is simulated numerically to investigate various parameters and support the presented formulation with the necessary graphical plots.

6.2 Preliminaries

We summarily furnish the requisites of fractional-order operators (see [7, 12]) to work out the suggested research problem. For a comprehensive study of fractional calculus, one may go through [11].

Definition 6.2.1. For a continuous function f defined over [a, b] and $\alpha > 0$, the αth-order Riemann–Liouville fractional integral is given by

$$_a I_t^\alpha f(t) = \frac{1}{\Gamma(\alpha)} \int_a^t (t-\tau)^{\alpha-1} f(\tau)\, d\tau.$$

Definition 6.2.2. For a continuous function f defined over [a, b], $\alpha > 0$ and $n-1 \le \alpha < n$, the αth-order Riemann–Liouville fractional derivative is given by

$$_a D_t^\alpha f(t) = \frac{1}{\Gamma(n-\alpha)} \left(\frac{d}{dt}\right)^n \int_a^t (t-\tau)^{n-\alpha-1} f(\tau)\, d\tau.$$

Definition 6.2.3. For a n-times continuously differentiable function f defined over [a, b] and $n-1 \le \alpha < n$, the αth-order Caputo fractional derivative is given by

$$\,^C_a D^\alpha_t f(t) = \frac{1}{\Gamma(n-\alpha)} \int_a^t (t-\tau)^{n-\alpha-1} f^{(n)}(\tau)\, d\tau.$$

Some Properties of Fractional-Order Operators:

$\Rightarrow$ Connection in Riemann–Liouville and Caputo fractional derivatives:

$$\,_a D^\alpha_t f(t) = \,^C_a D^\alpha_t f(t) + \sum_{k=0}^{n-1} \frac{(t-a)^{k-\alpha}}{\Gamma(k-\alpha+1)} f^{(k)}(\alpha).$$

$\Rightarrow$ Composition rules:

• $\,_a I^\alpha_t \,_a D^\alpha_t f(t) = f(t) - \sum_{j=0}^{n-1} \frac{(t-a)^{\alpha-j}}{\Gamma(\alpha-j+1)} \,_a D^{\alpha-j}_t f(a).$

• $\,_a I^\alpha_t \,^C_a D^\alpha_t f(t) = f(t) - \sum_{j=0}^{n-1} \frac{(t-a)^j}{\Gamma(j+1)} f^{(j)}(a).$

Definition 6.2.4. For $\alpha > 0$, the Mittag–Leffler function [33] , denoted by $E_\alpha(z)$, is outlined by an infinite series representation as

$$E_\alpha(z) = \sum_{r=0}^{\infty} \frac{z^r}{\Gamma(\alpha r+1)}.$$

6.3 Formulation of Fractional-Order Model of Hashimoto's Thyroiditis

We discuss a mathematical model of Hashimoto's Thyroiditis (HT) based on [32] by replacing the integer-order derivatives with α^{th}-order ($0 < \alpha < 1$) Caputo fractional derivative. The basic integer-order model (see [30], [34], [32], and references therein) describing the growth of autoimmune thyroiditis is given by

$$\frac{dTSH}{dt} = k_1 - \frac{k_1 FT4}{k_a+FT4} - k_2 TSH, \qquad TSH(t_0) = TSH_0,$$

$$\frac{dFT4}{dt} = \frac{k_3 T\, TSH}{k_d+TSH} - k_4 FT4, \qquad FT4(t_0) = FT4_0,$$

$$\frac{dT}{dt} = k_5 \left(\frac{TSH}{T} - N \right) - k_6 Ab\, T, \qquad T(t_0) = T_0,$$

$$\frac{dAb}{dt} = k_7\, Ab\, T - k_8 Ab, \qquad Ab(t_0) = Ab_0,$$

where the system dynamics of the above model involve four variables namely thyroid stimulating hormone (TSH), free thyroxin (FT4), anti-thyroid antibodies (Ab), and the functional size of the thyroid gland (T). This is one of the widely discussed classical 4D-Model with the associated initial conditions and the parameters k_1, k_2, k_3, k_4, k_5, k_6, k_7, k_8, k_a, k_d, N > 0. In order to avoid any complexities in the above stated model, we re-write the variables as

$$[TSH] = x_1, \quad [FT4] = x_2, \quad [T] = x_3, \quad \text{and} \quad [Ab] = x_4.$$

To introduce the effect of non-locality, we extend the classical model to a fractional model by employing the Caputo derivative as explained below:

$$_0^C D_t^\alpha x_1 = \frac{k_1 k_a}{k_a + x_2} - k_2 x_1, \tag{6.1}$$

$$_0^C D_t^\alpha x_2 = \frac{k_3 x_1 x_3}{k_d + x_1} - k_4 x_2, \tag{6.2}$$

$$_0^C D_t^\alpha x_3 = k_5 \left(\frac{x_1}{x_3} - N \right) - k_6 x_3 x_4, \tag{6.3}$$

$$_0^C D_t^\alpha x_4 = k_7 x_3 x_4 - k_8 x_4, \tag{6.4}$$

with the initial condition $x_i(0) = x_{i0}, 1 \le i \le 4$. We emphasize the proposed fractional-order model (1)–(4) as the fractional-order differential equations are instinctively connected to the memory preserving biological systems. The presence of fractional-order derivatives signifies that the presented model depends on the current as well as all the past states.

6.4 Stability Analysis

We now explain the local stability conditions of the equilibrium points of the fractional model (1)-(4) suggested in Section 3. Further, the steady states for this problem can be obtained by

$$_0^C D_t^\alpha x_4 = 0, \quad i = 1, 2, 3, 4 \quad \alpha \in (0, 1), \tag{6.5}$$

and the solutions of Eq. (6.5) represent the required equilibrium points, denoted by x_i^{eq} ($1 \le i \le 4$), of the problem (1)-(4). Next, to examine the local asymptotic stability, we write

$$x_i(t) = x_i^{eq} + \delta_i(t), \quad i = 1, 2, 3, 4,$$

which gives us

$$_0^C D_t^\alpha x_i^{eq} + \delta_i\,(t) = f_i\,(x_1^{eq} + \delta_1\,(t)\,,x_2^{eq} + \delta_2\,(t)\,,x_3^{eq} + \delta_3\,(t)\,,x_4^{eq} + \delta_4\,(t))\,.$$

By using $_0^C D_t^\alpha x_i^{eq} = 0$, we obtain

$$_0^C D_t^\alpha \delta_i\,(t) = f_i\,(x_1^{eq} + \delta_1\,(t)\,,x_2^{eq} + \delta_2\,(t)\,,x_3^{eq} + \delta_3\,(t)\,,x_4^{eq} + \delta_4\,(t))\,,$$

and we can further express f_i's as

$$f_i\,(x_1^{eq} + \delta_1\,(t)\,,x_2^{eq} + \delta_2\,(t)\,,x_3^{eq} + \delta_3\,(t)\,,x_4^{eq} + \delta_4\,(t)) \cong f_i\,(x_1^{eq},x_2^{eq},x_3^{eq},x_4^{eq}) +$$

$$\left[\delta_1 \frac{\partial}{\partial x_1}\,(f_i) + \delta_2 \frac{\partial}{\partial x_2}\,(f_i) + \delta_3 \frac{\partial}{\partial x_3}\,(f_i) + \delta_4 \frac{\partial}{\partial x_4}\,(f_i)\right]_{(x_1^{eq},x_2^{eq},x_3^{eq},x_4^{eq})}.$$

Again, by using $f_i\,(x_1^{eq},x_2^{eq},x_3^{eq},x_4^{eq}) = 0$, we get

$$_0^C D_t^\alpha \delta_i\,(t) = \left[\delta_1 \frac{\partial}{\partial x_1}\,(f_i) + \delta_2 \frac{\partial}{\partial x_2}\,(f_i) + \delta_3 \frac{\partial}{\partial x_3}\,(f_i) + \delta_4 \frac{\partial}{\partial x_4}\,(f_i)\right]$$

$$_{(x_1^{eq},x_2^{eq},x_3^{eq},x_4^{eq})}.$$

In the equivalent matrix notation, we express the above system as

$$_0^C D_t^\alpha \overrightarrow{\delta}\,(t) = J\overrightarrow{\delta}\,(t)\,,\ \text{with}\ \ \delta_i\,(0) = x_i\,(0) - x_i^{eq},\ 1\leq i \leq 4,$$

Where

$$\overrightarrow{\delta}\,(t) = \begin{bmatrix} \delta_1 \\ \delta_2 \\ \delta_3 \\ \delta_4 \end{bmatrix} \ and$$

$$J = \begin{bmatrix} \frac{\partial}{\partial x_1}\,(f_1) & \frac{\partial}{\partial x_2}\,(f_1) & \frac{\partial}{\partial x_3}\,(f_1) & \frac{\partial}{\partial x_4}\,(f_1) \\ \frac{\partial}{\partial x_1}\,(f_2) & \frac{\partial}{\partial x_2}\,(f_2) & \frac{\partial}{\partial x_3}\,(f_2) & \frac{\partial}{\partial x_4}\,(f_2) \\ \frac{\partial}{\partial x_1}\,(f_3) & \frac{\partial}{\partial x_2}\,(f_3) & \frac{\partial}{\partial x_3}\,(f_3) & \frac{\partial}{\partial x_4}\,(f_3) \\ \frac{\partial}{\partial x_1}\,(f_4) & \frac{\partial}{\partial x_2}\,(f_4) & \frac{\partial}{\partial x_3}\,(f_4) & \frac{\partial}{\partial x_4}\,(f_4) \end{bmatrix}_{(x_1^{eq},x_2^{eq},x_3^{eq},x_4^{eq})}$$

Or, in simpler notation, we can write

$$_0^C D_t^\alpha,\ \text{with}\ \delta_i\,(0) = x_i\,(0) - x_i^{eq},\ 1\leq i \leq 4$$

with $\overrightarrow{\delta} = (\delta_i)_{1\leq i \leq 4}, J = (J_{ij})_{1\leq i \leq 4},$ where $J_{ij} = \dfrac{\partial f_i}{\partial x_j}\bigg|_{(x_1^{eq},x_2^{eq},x_3^{eq},x_4^{eq})}.$

Assume that λ_i's $(1 \leq i \leq 4)$ are the eigenvalues of the matrix J, and B is the matrix of eigenvectors corresponding to the eigenvalues of J, so that we have

$$B^{-1}JB = \lambda, \text{ where } \lambda = diag\ (\lambda_1, \lambda_2, \lambda_3, \lambda_4).$$

Or,

$$J = B\Lambda B^{-1}$$

$$\Rightarrow {}_0^C D_t^\alpha \vec{\delta} = \left(B\Lambda B^{-1}\right) \vec{\delta}$$

$$\Rightarrow {}_0^C D_t^\alpha \left(B^{-1}\vec{\delta}\right) = \Lambda \left(B^{-1}\vec{\delta}\right),$$

$$\Rightarrow {}_0^C D_t^\alpha \left(\vec{\beta}\right) = \Lambda \vec{\beta}, \text{ where } \vec{\beta} = B^{-1}\vec{\delta} = (\beta_i)_{1 \leq i \leq 4},$$

and the solution to the above α^{th}-order differential equation (see [35]) is

$$\beta_i = E_\alpha \left(\lambda_i t^\alpha\right) \beta_i\ (0) \qquad 1 \leq i \leq 4.$$

Further the equilibrium point x_i^{eq} is asymptotically locally stable if all the eigenvalues of Jacobian matrix

$$J = \begin{bmatrix} \frac{\partial}{\partial x_1}(f_1) & \frac{\partial}{\partial x_2}(f_1) & \frac{\partial}{\partial x_3}(f_1) & \frac{\partial}{\partial x_4}(f_1) \\ \frac{\partial}{\partial x_1}(f_2) & \frac{\partial}{\partial x_2}(f_2) & \frac{\partial}{\partial x_3}(f_2) & \frac{\partial}{\partial x_4}(f_2) \\ \frac{\partial}{\partial x_1}(f_3) & \frac{\partial}{\partial x_2}(f_3) & \frac{\partial}{\partial x_3}(f_3) & \frac{\partial}{\partial x_4}(f_3) \\ \frac{\partial}{\partial x_1}(f_4) & \frac{\partial}{\partial x_2}(f_4) & \frac{\partial}{\partial x_3}(f_4) & \frac{\partial}{\partial x_4}(f_4) \end{bmatrix}$$

evaluated at the equilibrium point x_i^{eq} satisfies $|arg\ (\lambda_i)| > \frac{\alpha\pi}{2}$ (see [36]).
On simplifying $J = (J_{ij})_{1 \leq i,j \leq n}$, $J_{ij} = \frac{\partial f_i}{\partial x_j}$, we get

$$J = \begin{bmatrix} -k_2 & -\frac{k_1 k_a}{(k_a+x_2)^2} & 0 & 0 \\ \frac{k_d k_3 x_3}{(k_a+x_1)^2} & -k_4 & \frac{k_3 x_1}{(k_d+x_1)} & 0 \\ \frac{k_5}{x_3} & 0 & -\frac{k_5 x_1}{x_3^2} - k_6 x_4 & -k_6 x_3 \\ 0 & 0 & k_7 x_4 & k_7 x_3 - k_8 \end{bmatrix}$$

$$(6.6)$$

On solving $f_i = 0$ $(1 \leq i \leq 4)$ for equilibrium points, we arrive at:

Case-I

$$x_1 = \frac{k_1}{k_2} \frac{k_a}{(k_a+x_2)}, \qquad (6.7)$$

$$x_2 = \frac{k_3}{k_4} \frac{x_1 x_3}{(k_d + x_1)}, \tag{6.8}$$

$$x_3 = \frac{x_1}{N}, \tag{6.9}$$

$$x_4 = 0. \tag{6.10}$$

And, the Jacobian matrix given by Eq. (6) reduces to

$$J = \begin{bmatrix} -k_2 & -\frac{k_1 k_a}{(k_a + x_2)^2} & 0 & 0 \\ \frac{k_d k_3 x_1}{N(k_a + x_1)^2} & -k_4 & \frac{k_3 x_1}{(k_d + x_1)} & 0 \\ \frac{k_5 N}{x_3} & 0 & -\frac{k_5 N^2}{x_1} & -k_6 x_1 \\ 0 & 0 & 0 & \frac{k_7 x_1}{N} - k_8 \end{bmatrix},$$

where x_1 and x_2 are related by the Eqs. (6.7) and (6.8). Further, the variable x_1 can be obtained by solving the following cubic equation

$$k_2 k_3 x_1^3 + N k_2 k_a k_4 x_1^2 + (N k_2 k_a k_4 k_d - N k_1 k_a k_4) x_1 - N k_1 k_a k_4 k_d = 0.$$

Case:II

$$x_1 = \frac{k_1}{k_2} \frac{k_a}{(k_a + x_2)}, \tag{6.11}$$

$$x_2 = \frac{k_3}{k_4} \frac{k_8}{k_7} \frac{x_1 x_3}{(k_d + x_1)}, \tag{6.12}$$

$$x_3 = \frac{k_8}{k_7}, \tag{6.13}$$

$$x_4 = \frac{k_5}{k_6} \frac{x_1 - N x_3}{x_3^2}. \tag{6.14}$$

Again, the Jacobian matrix given by Eq. (6.6) reduces to

$$J = \begin{bmatrix} -k_2 & -\frac{k_1 k_a}{(k_a + x_2)^2} & 0 & 0 \\ \frac{k_d k_3 x_8}{x_7(k_a + x_1)^2} & -k_4 & \frac{k_3 x_1}{(k_d + x_1)} & 0 \\ \frac{k_5 k_7}{k_8} & 0 & -\frac{k_5 x_1 k_7^2}{k_8^2} - \frac{k_7 k_5 (k_7 x_1 - N x_8)}{k_8^2} & -k_6 \frac{k_7}{k_8} \\ 0 & 0 & \frac{k_5 k_7^2 (k_7 x_1 - N x_8)}{k_8^2} & 0 \end{bmatrix},$$

where x_1 and x_2 are related by the Eqs. (6.11) and (6.12). Further, the variable x_1 can be obtained by solving the following quadratic equation

$$(k_2 k_3 k_8 + k_2 k_a k_4 k_7) x_1^2 + k_7 k_a k_4 (k_2 k_d - k_1) x_1 - k_1 k_a k_4 k_d k_7 = 0.$$

6.5 Construction of a Numerical Solution Scheme

Next, we implement the Adomian decomposition method (ADM) (see [37,38,39], and references therein) to develop an approximate solution scheme of the problem (1)-(4), represented by a set of four α^{th}-order differential equations. To reduce the complexity in working process, we rewrite the problem (1)-(4)

$$_0^C D_t^\alpha x_i = f_i, \quad 1 \leq i \leq 4, \tag{6.15}$$

where

$$f_1 = \frac{k_1 k_a}{(k_a + x_2)} - k_2 x_1,$$

$$f_2 = \frac{k_3 x_1 x_3}{(k_d + x_1)} - k_4 x_2,$$

$$f_3 = k_5 \left(\frac{x_1}{x_3} - N \right) - k_6 x_3 x_4,$$

$$f_4 = k_7 x_3 x_4 + k_8 x_4.$$

To apply ADM, we express x_i's $(1 \leq i \leq 4)$ as an infinite series given by

$$x_i = \sum_{m=0}^{\infty} x_{im}, \quad 1 \leq i \leq 4.$$

On operating the αth-order Riemann–Liouville integral to Eq. (15), we obtain

$$x_i = x_i(0) + {}_0 I_t^\alpha f_i, \quad 1 \leq i \leq 4,$$

where $x_i(0) = x_{i0}$ is the initial state value. And, non-linear terms $f_i's$ in the Eq. (15) are expressed as

$$f_i = \sum_{m=0}^{\infty} A_{im}, \quad 1 \leq i \leq 4,$$

where $A_{im}'s$ are the Adomian polynomials given by

$$A_{im} = \frac{1}{m!} \left[\frac{\partial^m}{\partial \epsilon^m} f_{i\epsilon} \right]_{\epsilon=0},$$

and

$$f_{i\epsilon} = f_i(x_{1\epsilon}, x_{2\epsilon} \, x_{3\epsilon}, x_{4\epsilon}).$$

Thus, we arrive at the recurrence relation for x_i as

$$x_{i,m+1} = {}_0 I_t^\alpha A_{im}, \quad 1 \le i \le 4, \tag{6.16}$$

with the initial condition $x_{i0} = x_i(0)$. At last, we arrive at the following k^{th}-order approximate solution of the problem obtained by implementing ADM on Eq. (6.15) as

$$x_i = \lim_{k \to \infty} x_{ik},$$

where

$$x_{ik} = \sum_{m=0}^{k-1} x_{im}, \quad 1 \le i \le 4.$$

6.6 Numerical Segment

This section intends to execute the numerical simulations required to implement the ADM on the posed problem (6.1)–(6.4) (as discussed in Section 5). We request the readers to refer to Tables 6.1-6.2 for the estimated and normal range of parameter values (see [34, 32]) used in this section. One may mark that the Mathematica software has been used to perform the computations and plot the appropriate graphs throughout the work.

Table 6.1   Estimated values of parameters

Parameter	Estimated value	Unit
k_1	5000	$\frac{mU}{L*day}$
k_2	16.6355	$\frac{1}{day}$
k_3	100	$\frac{pg}{mL*L*day}$
k_4	0.099021	$\frac{1}{day}$
k_5	1	$\frac{L^3}{mU*day}$
k_6	1	$\frac{mL}{U*day}$
k_7	2.44	$\frac{1}{L*day}$
k_8	0.035	$\frac{1}{day}$
k_a	0.039	$\frac{pg}{mL}$
k_d	0.0083	$\frac{mU}{L}$
N	58.516	$\frac{mU}{L^2}$

Table 6.2 Standard values and range of the variables $x_i, 1 \leq i \leq 4$.

Name	Variable	Standard Value	Standard Range	Unit
TSH	x_1	1	0.4-2.5	$\frac{mU}{L}$
FT4	x_2	13	7-18	$\frac{pg}{mL}$
T	x_3	0.015	0.005-0.125	L
Ab	x_4	0	0-200	$\frac{U}{mL}$

By using the parameter values listed in Table 6.1 and the initial values

$$x_1(0) = 0.828 \, , \, x_2(0) = 14.15 \, , \, x_3(0) = 0.015 \, , \, x_4(0) = 50.92,$$

the Adomian polynomials evaluated by Mathematica are given as follows:

$$A_{10} = -0.0311536, \quad A_{20} = 0.0839658,$$

$$A_{30} = -4.0798, \quad A_{40} = 0.081472,$$

$$A_{11} = 0.436929 \frac{t^\alpha}{\Gamma(1+\alpha)}, \quad A_{21} = -403.94 \frac{t^\alpha}{\Gamma(1+\alpha)},$$

$$A_{31} = 15219.3 \frac{t^\alpha}{\Gamma(1+\alpha)}, \quad A_{41} = -506.894 \frac{t^\alpha}{\Gamma(1+\alpha)},$$

$$A_{12} = 0.000481265 \frac{t^{2\alpha}}{\Gamma(1+\alpha)^2} + 383.975 \frac{t^{2\alpha}}{\Gamma(1+\alpha)}$$

$$A_{22} = 30.2492 \frac{t^{2\alpha}}{\Gamma(1+\alpha)^2} + 1.50687*10^6 \frac{t^{2\alpha}}{\Gamma(1+\alpha)},$$

$$A_{32} = 4.08408*10^6 \frac{t^{2\alpha}}{\Gamma(1+\alpha)^2} - 5.67821*10^7 \frac{t^{2\alpha}}{\Gamma(1+\alpha)^2},$$

$$A_{42} = -0.81103 \frac{t^{2\alpha}}{\Gamma(1+\alpha)^2} - 1.89092*10^6 \frac{t^{2\alpha}}{\Gamma(1+2\alpha)}.$$

And, the recurrence relation (6.16) explained in Section 5 induces the following numerical scheme

$$x_{1,m+1} = {}_0 I_t^\alpha A_{1m}, \quad x_{10} = x_1(0) = 0.828,$$

$$x_{2,m+1} = {}_0 I_t^\alpha A_{2m}, \quad x_{20} = x_2(0) = 14.15,$$

$$x_{3,m+1} = {}_0 I_t^\alpha A_{3m}, \quad x_{30} = x_3(0) = 0.015,$$

$$x_{4,m+1} = {}_0I_t^\alpha A_{4m}, \; x_{40} = x_4\,(0) = 50.92.$$

In the first iteration (m=0):

$$x_{11} = {}_0I_t^\alpha A_{10} = -0.0311536 \; \frac{t^\alpha}{\Gamma\,(1+\alpha)},$$

$$x_{21} = {}_0I_t^\alpha A_{20} = 0.0839658 \; \frac{t^\alpha}{\Gamma\,(1+\alpha)},$$

$$x_{31} = {}_0I_t^\alpha A_{30} = -4.0798 \; \frac{t^\alpha}{\Gamma\,(1+\alpha)},$$

$$x_{41} = {}_0I_t^\alpha A_{40} = 0.081472 \; \frac{t^\alpha}{\Gamma\,(1+\alpha)}.$$

In the second iteration (m=1):

$$x_{12} = {}_0I_t^\alpha A_{11} = 0.436929 \; \frac{t^{2\alpha}}{\Gamma\,(1+2\alpha)},$$

$$x_{22} = {}_0I_t^\alpha A_{21} = -403.94 \; \frac{t^\alpha}{\Gamma\,(1+2\alpha)},$$

$$x_{32} = {}_0I_t^\alpha A_{31} = 15219.3 \; \frac{t^\alpha}{\Gamma\,(1+2\alpha)},$$

$$x_{42} = {}_0I_t^\alpha A_{41} = -506.894 \; \frac{t^\alpha}{\Gamma\,(1+2\alpha)}.$$

In the third iteration (m=2):

$$x_{13} = {}_0I_t^\alpha A_{12} = 0.000481265 \frac{\Gamma\,(1+2\alpha)\,t^{3\alpha}}{\Gamma\,(1+3\alpha)\,\Gamma(1+\alpha)^2} + 383.975 \; \frac{t^{3\alpha}}{\Gamma\,(1+3\alpha)},$$

$$x_{23} = {}_0I_t^\alpha A_{22} = 30.2492 \; \frac{\Gamma\,(1+2\alpha)\,t^{3\alpha}}{\Gamma\,(1+3\alpha)\,\Gamma(1+\alpha)^2} + 1.50687*10^6 \; \frac{t^{3\alpha}}{\Gamma\,(1+3\alpha)},$$

$$x_{33} = {}_0I_t^\alpha A_{32} = 4.08408*10^6 \; \frac{\Gamma\,(1+2\alpha)\,t^{3\alpha}}{\Gamma\,(1+3\alpha)\,\Gamma(1+\alpha)^2}$$

$$-5.67821*10^7 \; \frac{t^{3\alpha}}{\Gamma\,(1+3\alpha)},$$

$$x_{43} = {}_0I_t^\alpha A_{42} = -0.81103 \; \frac{\Gamma\,(1+2\alpha)\,t^{3\alpha}}{\Gamma\,(1+3\alpha)\,\Gamma(1+\alpha)^2} - 1.89092*10^6 \; \frac{t^{2\alpha}}{\Gamma\,(1+2\alpha)}.$$

Using the above iterations, we can write

$$x_1(t) = x_{10}(t) + x_{11}(t) + x_{12}(t) + x_{13}(t) + \ldots,$$

$$x_2(t) = x_{20}(t) + x_{21}(t) + x_{22}(t) + x_{23}(t) + \ldots,$$

$$x_3(t) = x_{30}(t) + x_{31}(t) + x_{32}(t) + x_{33}(t) + \ldots,$$

$$x_4(t) = x_{40}(t) + x_{41}(t) + x_{42}(t) + x_{43}(t) + \ldots,$$

and thus variable $x_i(t)$ can be approximated (3rd -order) as

$$TSH(t) = x_1(t) \cong x_{10}(t) + x_{11}(t) + x_{12}(t) + x_{13}(t),$$

$$= 0.828 - 0.0311536 \frac{t^\alpha}{\Gamma(1+\alpha)} + 0.436929 \frac{t^{2\alpha}}{\Gamma(1+2\alpha)}$$

$$+ 0.000481265 \frac{\Gamma(1+2\alpha) t^{3\alpha}}{\Gamma(1+3\alpha) \Gamma(1+\alpha)^2} + 383.975 \frac{t^{3\alpha}}{\Gamma(1+3\alpha)},$$

$$FT4(t) = x_1(t) \cong x_{20}(t) + x_{21}(t) + x_{22}(t) + x_{23}(t),$$

$$= 14.15 + 0.0839658 \frac{t^\alpha}{\Gamma(1+\alpha)} - 403.94 \frac{t^\alpha}{\Gamma(1+2\alpha)}$$

$$+ 30.2492 \frac{\Gamma(1+2\alpha) t^{3\alpha}}{\Gamma(1+3\alpha) \Gamma(1+\alpha)^2} + 1.50687*10^6 \frac{t^{3\alpha}}{\Gamma(1+3\alpha)},$$

$$T(t) = x_3(t) \cong x_{30}(t) + x_{31}(t) + x_{32}(t) + x_{33}(t),$$

$$= 0.015 - 4.0798 \frac{t^\alpha}{\Gamma(1+\alpha)} + 15219.3 \frac{t^\alpha}{\Gamma(1+2\alpha)}$$

$$+ 4.08408 * 10^6 \frac{\Gamma(1+2\alpha) t^{3\alpha}}{\Gamma(1+3\alpha) \Gamma(1+\alpha)^2} - 5.67821*10^7 \frac{t^{3\alpha}}{\Gamma(1+3\alpha)},$$

$$Ab(t) = x_4(t) \cong x_{40}(t) + x_{41}(t) + x_{42}(t) + x_{43}(t),$$

$$= 50.92 + 0.081472 \frac{t^\alpha}{\Gamma(1+\alpha)} - 506.894 \frac{t^\alpha}{\Gamma(1+2\alpha)}$$

$$- 0.81103 \frac{\Gamma(1+2\alpha) t^{3\alpha}}{\Gamma(1+3\alpha) \Gamma(1+\alpha)^2} - 1.89092*10^6 \frac{t^{2\alpha}}{\Gamma(1+2\alpha)}.$$

With the aid of the third-order approximation computed above, we observe the plots of time-dependent variables TSH, FT4, T, Ab in Figures 6.1-6.2, for $\alpha \in (0, 1)$. Note that the notations TSH(t, α), FT4(t, α), Ab(t, α), T(t, α) marked in these Figures 6.1-6.4 represent the corresponding variables x_i ($1 \leq i \leq 4$) as a function of time t, for various fractional values of α.

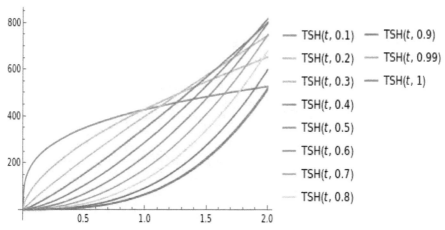

Figure 6.1   Plot of the variable $x_1(t, \alpha) = $ TSH(t, α) (Thyroid Stimulating Harmone).

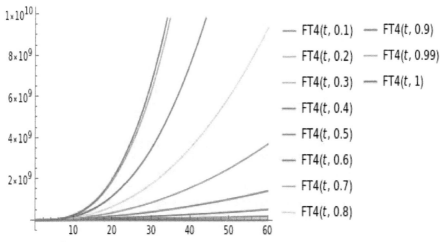

Figure 6.2   Plot of the variable $x_2(t, \alpha) = $ FT 4(t, α) (Free Thyroxin).

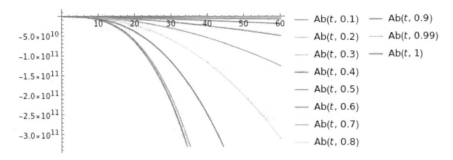

Figure 6.3 Plot of the variable $x_3(t, \alpha) = Ab(t, \alpha)$ (Anti-Thyroid Antibodies).

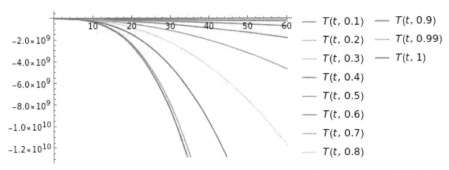

Figure 6.4 Plot of the variable $x_4(t, \alpha) = T(t, \alpha)$ (Functional Size of the Thyroid Gland).

6.7 Conclusions

We analyzed a fractional-order 4-d mathematical model of an auto-immune disease, Hashimoto's thyroiditis. Apart from describing the evident relevance of formulating a fractional-order model, we focused on explaining the local stability conditions. Further, we exercised the ADM to acquire the numerically simulated solutions to the posed problem. We firmly believe that the implications of our results demonstrated with necessary graphical interpretations shall attract the researchers to adopt non-integer order models more frequently that can handle the memory-preserving bio-mathematical models effectively.

References

[1] D. Kirschner, J. Panetta, Modelling immunotherapy of the tumor-immune interaction, Journal of Mathematical Biology 37 (1998) 235–252.

[2] T. Sardar, S. Rana, J. Chattopadhyay, A mathematical model of dengue transmission with memory, Communications in Nonlinear Science and Numerical Simulation 22 (2015) 511–525.

[3] N. Chitnis, J. Hyman, J. Cushing, Determining important parameters in the spread of malaria through the sensitivity analysis of a mathematical model, Bulletin of Mathematical Biology 70 (2008) 1272–1296.

[4] A. Huppert, G. Katriel, Mathematical modelling and prediction in infectious disease epidemiology, Clinical Microbiology and Infection 19 (2013) 999–1005.

[5] J. Huo, H. Zhao, Dynamical analysis of a fractional sir model with birth and death on heterogeneous complex networks, Physica A: Statistical Mechanics and its Applications 448 (2016) 41–56.

[6] I. Podlubny, Geometric and physical interpretation of fractional integration and fractional differentiation, Fractional Calculus and Applied Analysis 5 (2002) 367–386.

[7] M. Moshrefi-Torbati, J. Hammond, Physical and geometrical interpretation of fractional operators, Journal of the Franklin Institute 335 (1998) 1077–1086.

[8] V. Tarasov, Geometric interpretation of fractional-order derivative, Fractional Calculus and Applied Analysis 19 (2016) 1200–1221.

[9] H. Joshi, B. Jha, On a reaction–diffusion model for calcium dynamics in neurons with Mittag–Leffler memory, Eur. Phys. J. Plus 136.

[10] K. Oldham, J. Spanier, The fractional calculus: Theory and applications of differentiation and integration of arbitrary order, Academic Press, New York.

[11] K. Miller, B. Ross, An introduction to the fractional calculus and fractional differential equations, John Wiley and Sons, New York.

[12] G. Zaslavsky, Chaos, fractional kinetics, and anomalous transport, Physics Reports 371 (2002) 461–580.

[13] S. Yuste, L. Acedo, K. Lindenberg, Reaction front in an A + B → C reaction-subdiffusion process, Physical Review E 69, Article ID 036126.

[14] H. Sheng, Y. Chen, T. Qiu, Fractional processes and Fractional-Order signal processing, Springer, New York, NY, USA.

[15] K. Assaleh, W. Ahmad, Modeling of speech signals using fractional calcu lus, in Proceedings of the 9th International Symposium on Signal Processing and its Applications (ISSPA '07), Sharjah, United Arab Emirates.

[16] H. Xu, Analytical approximations for a population growth model with fractional order, Communications in Nonlinear Science and Numerical Simulation 14 (2009) 1978–1983.

[17] L. Debnath, Recent applications of fractional calculus to science and engineering, International Journal of Mathematics and Mathematical Sciences 54 (2003) 3413–3442.

[18] R. Yafia, Hopf bifurcation in differential equations with delay for tumor-immune system competition model, SIAM Journal on Applied Mathematics 67 (2007) 1693–1703.

[19] H. Joshi, B. Jha, Fractional-order mathematical model for calcium distribution in nerve cells, Comp. Appl. Math 39.

[20] E. Ahmed, A. Hashish, F. Rihan, On fractional order cancer model, Journal of Fractional Calculus and Applied Analysis 3 (2012) 1–6.

[21] M. Dokuyucu, E. Celik, H. Bulut, H. Baskonus, Cancer treatment model with the Caputo-Fabrizio fractional derivative, The European Phys. J. Plus 133.

[22] N. Singha, C. Nahak, A numerical scheme for generalized fractional optimal control problems, Appl. Appl. Math. 11 (2016) 798–814.

[23] N. Singha, Implementation of fractional optimal control problems in real- world applications, Fractional Calculus and Applied Analysis 23 (2020) 1783–1796.

[24] N. Singha, C. Nahak, An efficient approximation technique for solving a class of fractional optimal control problems, Journal of Optimization Theory and Applications 174 (2017) 785–802.

[25] H. Joshi, B. Jha, Modeling the spatiotemporal intracellular calcium dynamics in nerve cell with strong memory effects, International Journal of Nonlinear Sciences and Numerical Simulation.

[26] B. Pandiyan, S. Merrill, A model of the cost of delaying treatment of Hashimoto's thyroiditis: thyroid cancer initiation and growth, Mathematical Biosciences and Engineering 16 (2019) 8069–8091.

[27] F. Mainardi, R. Gorenflo, On Mittag–Leffler type functions in fractional evolution processes, J. Comput. Appl. Math. 118 (2010) 283–299.

[28] B. Pandiyan, S. Merrill, S. Benvenga, A patient-specific model of the negative-feedback control of the hypothalamus-pituitary-thyroid (HPT) axis in autoimmune (Hashimoto's) thyroiditis, Mathematical Biosciences and Engineering 31 (2013) 226–258.

[29] E. Ahmed, A. Elgazzar, On fractional order differential equations model for nonlocal epidemics, Physica A: Statistical Mechanics and its Applications 379 (2007) 607–614.

[30] G. Adomian, A review of the decomposition method and some recent results for nonlinear equation, Math. Comput. Modelling 13 (1990) 17–43.

[31] G. Adomian, D. Sarafyan, Numerical solution of differential equations in the deterministic limit to stochastic theory, Appl. Math. Comput. 8 (1981) 111–119.

7

Heated Laminar Vertical Jet of Pseudoplastic Fluids-Against Gravity

Manisha Patel[1] and M. G. Timol[2]

[1]Department of Mathematics, Sarvajanik College of Engineering &
Technology, Surat-395001, Gujarat, India
[2]Retired Professor, Department of Mathematics, Veer Narmad South Gujarat
University, Magdalla Road, Surat-395007, Gujarat, India
E-mail: manishapramitpatel@gmail.com; mgtimol@gmail.com

Abstract

A heated laminar jet of Pseudo-plastic fluid flowing vertically upwards from
a long narrow slit into a region of the same fluid at rest and at a uniform temperature is considered. The governing non-linear Partial differential equations
(PDEs) for the defined flow problem are transformed into non-linear ordinary
differential equations using the effective similarity technique-one parameter
deductive group theory method. The obtained non-linear coupled Ordinary
differential equations are solved, and graphs present the results. The effect of
the Prandtl number and Grashof number on the velocity and temperature of
the jet flow is discussed.

Keywords: Laminar, vertical jet, Prandtl number, Grashof number, Pseudoplastic fluid, non-linear PDE.

7.1 Introduction

The theory of jet is an important and highly developed branch of hydromechanics. The Garden fountain is an example of the vertical jet of a Newtonian
fluid, shown in Figure 7.1. Lava (magma) is immersed in a vent as a fire

Figure 7.1   Garden Fountain.

Figure 7.2   Volcanic Irruption.

fountain and can be considered a huge natural vertical jet of non-Newtonian fluid, as given in Figure 7.2. The theory of free and confined jets has applications in cardiac blood flow-related problems. Jet flows are consequences of many cardiac lesions. As blood is the Pseudoplastic fluid, the blood flow through the heart can also be considered an important example of the vertical jet flow of Pseudoplastic fluid, see Figure 7.3.

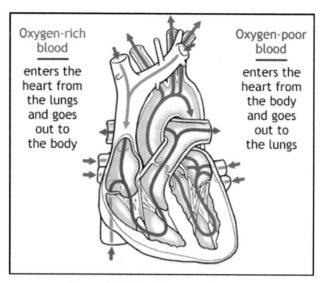

Figure 7.3   Blood flow through heart.

The first problem of jet for ideal fluid was solved by Helmholtz (1868). Kirchhoff (1869) substantially developed and generalized Helmholtz's method. Potential flow under gravity with free surfaces has been somewhat neglected in classical hydromechanics. Advanced technology requires more and more understanding of the problems of jet flows. There are numerous situations in aerodynamics, engineering processes, and meteorology where jet flows occur in a natural way. Consider a fluid moving in a tank or pipe. It is observed that as it crosses a slit or a hole, there is a sudden decrease in a cross-sectional area, and consequently, there is a considerable increase in velocity. It gives rise to the flow of a jet. Fluid flows in the absence of rigid boundaries, and therefore it is a free flow. A lot of work has been carried out in literature on the jet flows of compressible and incompressible fluids.

Schlichting (1933, 1968) was the first to expand the boundary layer theory to the theory of jet flows. The numerical solution of the governing ordinary differential equation of the steady two-dimensional free jet flow was determined by him. An analytic solution to this problem was discussed by Bickely (1937). H. Stehr *et al.* (2000) discussed the resulting effects because of the interaction of the induced flow and the jet. They have applied both numerical and analytical methods for the solution of the governing equations. The model and computation of an inviscid liquid jet emerging in a rotating drum were analyzed by Decent *et al.* (2002). A slender liquid jet immersed from a liquid zone due to a circulating orifice was investigated by Wallwork *et al.* (2002). The author developed the model for the slender nonlinear inviscid jet with the assumption of the stationary centreline of the jet and in the presence of surface tension. Patel M *et al.* (2014) investigated the axisymmetrical and two-dimensional jet flow for an incompressible Pseudo-plastic fluid. They have considered the horizontal jet flow problem to obtain the similarity for the governing Partial differential equations. The jet flow of a laminar compressible electrically conducting fluid issuing from circular orifices in the presence of a radial magnetic field is analyzed by Manisha et al. (2015). The laminar jet flow due to an incompressible Newtonian fluid coming out from a circular hole or a narrow slit was investigated by M Patel *et al.* (2016). They have discussed both the cases, Free jet, and dissipative jet. They considered the variable physical properties: thermal conductivity and viscosity vary with temperature. Magan *et al.* (2016) analyzed the free jet of power-law fluid for which the Reynold number is considered in terms of the viscosity of the power-law fluid. The free jet is modeled by making the boundary layer approximation perpendicular to the axis of symmetry. Danial Riahi (2017) developed the governing equations and the boundary conditions for the flow of the nonlinear rotating slender jet. He then obtained the time-independent nonlinear solution to the problem.

Nowadays, the study of Pseudoplastic fluid (shear-thinning fluids) is essential because the fluids' important applications fall in this category. Drilling fluids, polymer solutions, Blood, Catch up, and Nail polish are examples of Pseudoplastic behavior. While drilling the boreholes into the surface of the earth or drilling the oil or natural gases, the purpose of Drilling fluids (mud) is to clean the bottom of the hole, lubricate the bit, maintain the walls of the hole, and transport the cuttings to the surface. This type of behavior is characteristic of suspensions of asymmetric particles or solutions of a high polymer such as cellulose derivatives. For these types of fluids, the apparent viscosity continues to decrease with increasing shear rate until no further alignment along the streamlines is possible, and the flow curve becomes linear.

An empirical functional relation known as Power-law is widely used to characterize fluids of this type. This relation can be written as

$$\tau = k\,\theta^n.$$

Where k and n are constants (n < 1) for the particular fluid: k is a measure of the consistency of the fluids, the higher k more viscous the fluid; n is the measure of the degree of non-Newtonian behavior. Other empirical relations which have been used to describe Pseudoplastic behavior are Prandtl, Eyring, Powell-Eyring, Williamson, etc. (Wilkinson W. L. 1960). Derivation of the governing equations of a Pseudoplastic fluid was first provided by Schowalter (1960). Acrivos *et al.* (1960) obtained the numerical analysis of the boundary layer equations of both Pseudoplastic fluids (Shear-thinning) and Dilatant fluids (shear-thickening). A brief instruction and classification of various fluid models of non-Newtonian fluid are given in detail by Manisha et al. and Akshara et al. (2010, 2013).

The technique used in the present investigation is deductive group transformation, which leads to a similarity representation of the problem. In 1967 and 1968, Moran et al. (1967, 1968, 1968, 1968) presented a general systematic formalism for similarity analysis, where the given system of partial differential equations reduces to a system of ordinary differential equations. The details of the theory have been given in (Moran 1967, 1968, 1968, 1968). And another technique to find a similarity solution is the group-theoretic transformation, which converts partial differential equations into ordinary differential equations. The two-parameter group-theoretic transformation method had applied by Patil *et al.* (2012) and Darji *et al.* (2014). Recently, the deductive group transformation method was applied by many researchers successfully to various flow problems (Malek 2002, Parmar 2011, Adnan 2011, Darji 2013, Nita 2015, Patel 2017).

Very little work has been carried out for the heated vertical jet of non-Newtonian fluid in the past. Kalathia N. L. (1975) was probably the first to derive the governing equations for the heated vertical jet flow for power-law fluid. William Conway (1967) investigated the two-dimensional vertical jet for an irrotational steady flow of an incompressible inviscid fluid under gravity between two horizontal planes. Jordan *et al.* (1992) carried out a study of a laminar submerged jet of a Pseudoplastic fluid. They formed a laminar jet by inserting a polymer solution into a large chamber using a long tube. In the present work, we expanded the work of Kalathia N. L. (1975) by obtaining similar solutions and then numerical solutions along with the graphical presentation.

In this paper, we study a heated laminar jet of Pseudoplastic fluid flowing vertically upwards from a long narrow slit into a region of the same fluid which is at rest and at a uniform temperature. A constant pressure difference produces the flow through the slit. It is assumed that the flow is steady, laminar (i.e., having a small Reynolds number), and incompressible. The existence of temperature difference between the jet and its surroundings will give rise to buoyant forces. As soon as the jet sinks, it will be under the effect of these forces. The free convection velocity, variation in density, etc., will depend on these buoyant forces in a steady state.

For simplicity, the constant properties are postulated except for a small change in density due to temperature differences. Such simplified treatments do not appear unreasonable as long as the temperature difference involved is not too large. The coupled equation of motion is solved within the limitations of the physical aspects of the problem.

7.2 Basic Equations

Consider a heated laminar jet of Pseudoplastic fluid flowing vertically upwards from a long narrow slit into a region of a same fluid that is at a rest and a uniform temperature. The geometry of the flow is shown in the Figure 7.4. After neglecting viscous dissipation, the governing equations for such natural convection flow, following Kalathia *et al.* (1975), can be written as:

$$\frac{\partial \bar{u}}{\partial \bar{x}} + \frac{\partial \bar{v}}{\partial \bar{y}} = 0, \tag{7.1}$$

$$\bar{u}\frac{\partial \bar{u}}{\partial \bar{x}} + \bar{v}\frac{\partial \bar{u}}{\partial \bar{y}} = v\frac{\partial}{\partial \bar{y}}\left(\left|\frac{\partial \bar{u}}{\partial \bar{y}}\right|^{n-1}\frac{\partial \bar{u}}{\partial \bar{y}}\right) + g\beta\bar{\theta}, \quad 0 < n < 1, \tag{7.2}$$

$$\rho C_p\left(\bar{u}\frac{\partial \bar{\theta}}{\partial \bar{x}} + \bar{v}\frac{\partial \bar{\theta}}{\partial \bar{y}}\right) = K\frac{\partial^2 \bar{\theta}}{\partial \bar{y}^2}, \tag{7.3}$$

The necessary boundary conditions are

$$\bar{v} = \frac{\partial \bar{u}}{\partial \bar{y}} = \frac{\partial \bar{\theta}}{\partial \bar{y}} = 0 \quad at \quad \bar{y} = 0, \tag{7.4}$$

$$\bar{u} = \bar{\theta} = 0 \quad at \quad \bar{y} \to \infty, \tag{7.5}$$

It is to note that the usual condition of constancy of flux of the momentum will not be satisfied in the case of buoyant jets. However, following the assumptions made, we shall have (whether the jet is buoyant or not)

$$Q = \rho C_p \int_0^\infty \bar{u}\bar{\theta}\, d\bar{y} = C_1 = \text{Constant}, \tag{7.6}$$

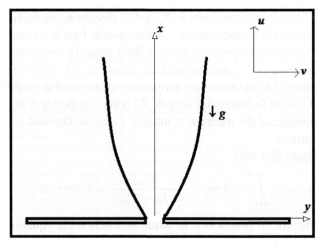

Figure 7.4 Co-ordinate system for two-dimensional vertical jet

Introducing the non-dimensional variable as follows

$$x = \frac{\bar{x}}{L}, \, y = \frac{\bar{y}}{L}\left(\frac{L^2}{v}\right)^{\frac{n-1}{n+1}}, \, u = \frac{L\bar{u}}{v}, \, v = \frac{L\bar{v}}{v}\left(\frac{L^2}{v}\right)^{\frac{n-1}{n+1}},$$

$$\theta = \frac{\bar{\theta}}{(T_0 - T_\infty)}, \tag{7.7}$$

The governing equations (7.1)-(7.6) in dimensionless form are given by equations (7.8)-(7.13):

$$\frac{\partial u}{\partial x} + \frac{\partial v}{\partial y} = 0, \tag{7.8}$$

$$u\frac{\partial u}{\partial x} + v\frac{\partial u}{\partial y} = \frac{v^{n-1}}{\alpha^n}\frac{\partial}{\partial y}\left(\left|\frac{\partial u}{\partial y}\right|^{n-1}\frac{\partial u}{\partial y}\right) + Gr\,\theta, \tag{7.9}$$

$$u\frac{\partial \theta}{\partial x} + v\frac{\partial \theta}{\partial y} = \frac{1}{Pr}\frac{\partial^2\theta}{\partial y^2}, \tag{7.10}$$

The boundary conditions are

$$v = \frac{\partial u}{\partial y} = \frac{\partial \theta}{\partial y} = 0 \text{ at } y = 0, \tag{7.11}$$

$$u = \theta = 0 \text{ at } y \to \infty, \tag{7.12}$$

u and v are the velocity components in X and Y directions respectively. ρ is the mass density, θ is the temperature, C_p is the specific heat at a constant temperature, K is the thermal conductivity, υ is the kinematic viscosity, n is the flow behavior index, g is the gravitational acceleration, β is the coefficient of thermal expansion, Q is the heat flux across any cross-section perpendicular to the jet axis. L is the fundamental length, T_0 is the temperature at the wall, T_∞ is the temperature of the medium at infinity, Gr is the Grashof number, Pr is the Prandtl number.

The momentum flux will

$$Q = \rho C_p \int_0^\infty u\theta \, dy = \frac{C_1}{\upsilon(T_0 - T_\infty)} = \text{Constant}, \qquad (7.13)$$

Introduce the stream function ψ as usual and hence the equations (7.8)-(7.13) come to the form

$$\frac{\partial\psi}{\partial y}\frac{\partial^2\psi}{\partial x\partial y} - \frac{\partial\psi}{\partial x}\frac{\partial^2\psi}{\partial y^2} = n\left(\frac{\partial^2\psi}{\partial y^2}\right)^{n-1}\frac{\partial^3\psi}{\partial y^3} + Gr\,\theta, \qquad (7.14)$$

$$\frac{\partial\psi}{\partial y}\frac{\partial\theta}{\partial x} - \frac{\partial\psi}{\partial x}\frac{\partial\theta}{\partial y} = \frac{1}{\text{Pr}}\frac{\partial^2\theta}{\partial y^2}, \qquad (7.15)$$

The boundary conditions are

$$\frac{\partial\psi}{\partial x} = \frac{\partial^2\psi}{\partial y^2} = \frac{\partial\theta}{\partial y} = 0 \quad at \quad y = 0, \qquad (7.16)$$

$$\frac{\partial\psi}{\partial y} = \theta = 0 \qquad\qquad at \quad y \to \infty, \qquad (7.17)$$

The momentum flux will

$$Q = \int_0^\infty \frac{\partial\psi}{\partial y}\theta \, dy \;\; = 1, \qquad (7.18)$$

To reduce the above equations ((7.14)-(7.18)) with two independent variables in the equations with one independent variable, we apply the one-parameter deductive group theory technique (Moran 1967, 1968, 1968, 1968).

Introducing the group G, (Patel 2017)

$$G: \begin{cases} \tilde{x} = h^x(a)\,x + k^x(a) \\ \tilde{y} = h^y(a)\,y + k^y(a) \\ \tilde{\psi} = h^\psi(a)\,\psi + k^\psi(a) \\ \tilde{\theta} = h^\theta(a)\,\theta + k^\theta(a) \end{cases}, \qquad (7.19)$$

Here, a is the group parameter. Equation (7.14) is said to be transformed invariantly for some function M(a), whenever,

$$\frac{\partial \tilde{\psi}}{\partial \tilde{y}} \frac{\partial^2 \tilde{\psi}}{\partial \tilde{x} \partial \tilde{y}} - \frac{\partial \tilde{\psi}}{\partial \tilde{x}} \frac{\partial^2 \tilde{\psi}}{\partial \tilde{y}^2} - n \left(\frac{\partial^2 \tilde{\psi}}{\partial \tilde{y}^2}\right)^{n-1} \frac{\partial^3 \tilde{\psi}}{\partial \tilde{y}^3} - Gr \, \tilde{\theta}$$

$$= M_1(a) \left[\frac{\partial \psi}{\partial y} \frac{\partial^2 \psi}{\partial x \partial y} - \frac{\partial \psi}{\partial x} \frac{\partial^2 \psi}{\partial y^2} - n \left(\frac{\partial^2 \psi}{\partial y^2}\right)^{n-1} \frac{\partial^3 \psi}{\partial y^3} - Gr \, \theta\right],$$

$$(7.20)$$

Similarly, we can say for the equation (7.15) to be absolutely invariant for some function M(a), whenever,

$$\frac{\partial \tilde{\psi}}{\partial \tilde{y}} \frac{\partial \tilde{\theta}}{\partial \tilde{x}} - \frac{\partial \tilde{\psi}}{\partial \tilde{x}} \frac{\partial \tilde{\theta}}{\partial \tilde{y}} - \frac{1}{Pr} \frac{\partial^2 \tilde{\theta}}{\partial \tilde{y}^2} = M_2(a) \left[\frac{\partial \psi}{\partial y} \frac{\partial \theta}{\partial x} - \frac{\partial \psi}{\partial x} \frac{\partial \theta}{\partial y} - \frac{1}{Pr} \frac{\partial^2 \theta}{\partial y^2}\right],$$

$$(7.21)$$

Therefore, from the above two equations (7.20) and (7.21), the following relations (7.22) and (7.23) are obtained respectively,

$$\frac{h^{2\psi}}{h^x h^{2y}} = \frac{h^{2\psi}}{h^x h^{2y}} = \left(\frac{h^\psi}{h^{2y}}\right)^{n-1} \frac{h^\psi}{h^{3y}} = h^\theta = M_1(a),$$

$$(7.22)$$

$$\frac{h^\psi h^\theta}{h^x h^y} = \frac{h^\theta}{h^{2y}} = M_2(a),$$

$$(7.23)$$

Also, from the invariance of the auxiliary conditions,

$$k^y = k^\theta = 0,$$

$$(7.24)$$

From the equations (7.22), (7.23) and (7.24)

$$h^x = h^{3y}, \quad h^\psi = h^{2y}, \quad h^\theta = h^{-y}, \quad k^\theta = k^y = 0,$$

$$(7.25)$$

Using (7.25) in (7.19), the group G is

$$G : \begin{cases} \tilde{x} = h^{3y}x + k^x \\ \tilde{y} = h^y y \\ \tilde{\psi} = h^{2y}\psi + k^\psi \\ \tilde{\theta} = h^{-1}\theta \end{cases},$$

$$(7.26)$$

$$\sum_{i=1}^{4} (\alpha_i S_i + \beta_i) \frac{\partial g}{\partial S_i} = 0; \quad S_i = x, y, \psi, \theta,$$

$$(7.27)$$

$$\Rightarrow (\alpha_1 x + \beta_1)\frac{\partial g}{\partial x} + (\alpha_2 y + \beta_2)\frac{\partial g}{\partial y} + (\alpha_3 \psi + \beta_3)\frac{\partial g}{\partial \psi} + (\alpha_4 \theta + \beta_4)\frac{\partial g}{\partial \theta} = 0,$$
$$(7.28)$$

Here

$$\alpha_i = \frac{\partial h^{S_i}}{\partial a} \quad \& \quad \beta_i = \frac{\partial k^{S_i}}{\partial a}; i = 1, 2, 3, 4, \qquad (7.29)$$

$$\alpha_1 = \frac{\partial h^x}{\partial a} = \frac{\partial h^{3y}}{\partial a} = 3h^{2y}\frac{\partial h^y}{\partial a} = 3\frac{\partial h^y}{\partial a} = 3\alpha_2, \quad (\because \ h^y \text{ is identity at } a_0)$$

$$\alpha_2 = \frac{\partial h^y}{\partial a}$$

$$\alpha_3 = \frac{\partial h^\psi}{\partial a} = \frac{\partial h^{2y}}{\partial a} = 2h^y\frac{\partial h^y}{\partial a} = 2\frac{\partial h^y}{\partial a} = 2\alpha_2, \quad (\because \ h^y \text{ is identity at } a_0)$$

$$\alpha_4 = \frac{\partial h^\theta}{\partial a} = \frac{\partial h^{-y}}{\partial a} = -h^{-2y}\frac{\partial h^y}{\partial a} = -\frac{\partial h^y}{\partial a} = -\alpha_2, \quad (\because \ h^y \text{ is identity at } a_0),$$
$$(7.30)$$

$$\beta_1 = \frac{\partial k^x}{\partial a}, \ \beta_2 = \frac{\partial k^y}{\partial a} = 0, \ \beta_3 = \frac{\partial k^\psi}{\partial a}, \ \beta_4 = \frac{\partial k^\theta}{\partial a} = 0, \qquad (7.31)$$

Now, the characteristic equation from the equation (7.28) is:

$$\frac{dx}{(\alpha_1 x + \beta_1)} = \frac{dy}{(\alpha_2 y + \beta_2)} = \frac{d\psi}{(\alpha_3 \psi + \beta_3)} = \frac{d\theta}{(\alpha_4 \theta + \beta_4)}, \qquad (7.32)$$

Solving first two relation of equation (7.32) for the similarity variable η, we have

$$\eta = y(\alpha_1 x + \beta_1)^{-\frac{1}{3}}, \qquad (7.33)$$

Solving first and third relation of equation (7.32) for the similarity function $f_1(\eta)$, we have

$$\psi = (\alpha_1 x + \beta_1)^{\frac{2}{3}} f_1(\eta) - \frac{\beta_3}{2\alpha_1}, \qquad (7.34)$$

Solving first and last relation of equation (7.32) for the similarity function $f_2(\eta)$, we have

$$\theta = (\alpha_1 x + \beta_1)^{-\frac{1}{3}} f_2(\eta), \qquad (7.35)$$

Using the equations (7.33), (7.34) and (7.35) and its derivatives in equations (7.14)-(7.18), the below similarity equations (7.36)-(7.40) are obtained.

$$n\left(\frac{d^2 f_1}{d\eta^2}\right)^{n-1}\frac{d^3 f_1}{d\eta^3} + \frac{2}{3}\alpha_1 f_1(\eta)\frac{d^2 f_1}{d\eta^2} - \frac{1}{3}\alpha_1\left(\frac{df_1}{d\eta}\right)^2 + Gr \ f_2(\eta) = 0,$$
$$(7.36)$$

$$\frac{1}{\mathrm{Pr}}\frac{d^2 f_2}{d\eta^2} + \frac{1}{3}\alpha_1 f_2(\eta)\frac{df_1}{d\eta} + \frac{2}{3}\alpha_1 f_1(\eta)\frac{df_2}{d\eta} = 0, \qquad (7.37)$$

$$f_1(\eta) = 0, \quad \frac{d^2 f_1}{d\eta^2} = 0, \quad \frac{df_2}{d\eta} = 0 \quad at \quad \eta = 0, \qquad (7.38)$$

$$\frac{df_1}{d\eta} = 0, \; f_2(\eta) = 0 \, at \, , \eta \to \infty, \qquad (7.39)$$

$$Q = \int_0^\infty f_2(\eta)\frac{df_1}{d\eta}\, d\eta = 1, \qquad (7.40)$$

7.3 Results and Discussions

The obtained similarity equations with auxiliary conditions (equations (7.36)-(7.40)) are solved using bvp4c- MATLAB ODE solver. The variation in velocity and temperature is present graphically. The analysis of the effect of the buoyant force and the gravitational forces for the temperature difference and velocity variation is very important for this type of problem. Figures 7.5, 7.6 & 7.7 present the graphs for the velocity profile for variation in the values of Prandtl number (Pr), Grashof number (Gr) and fluid index (n < 1), respectively. The velocity of jet flow remains constant for some interval of eta and then decrease rapidly with the increase in Prandtl number is shown by the graph in Figure 7.5. In other words, fluid is more viscous with a high Prandtl number, so velocity remains constant for a small interval and then decreases speedily. Gr & n fixed for this case. For the graph in Figure 7.6, Pr & n fixed, the velocity decreased more quickly with the increase in Gr. Particularly; the gravitational force effect is clearly shown in Figure 7.6 because at a high value of Gr, the fluid velocity is constant for a very small interval then it decreases quickly. Figure 7.7 represents the velocity decreased rapidly when the value of n decreases.

The variation in the temperature profile showed in Figures 7.8 & 7.9. Figure 7.8 represents the temperature profile for different values of Grashof number Gr. The wall temperature increase when the value of Gr or any buoyant related parameter increase. Because of this, the fluid particle bonding becomes weak and the internal friction decrease. And therefore, the gravity effects become strong. From Figure 7.8, the temperature increased rapidly with increases in the value of Gr, reached its peak value and then decreased briskly. Similarly, the temperature went quickly to its peak and then uniformly went down with the increase in the value of Pr, which is given in Figure 7.9.

7.4 Graphical Presentation

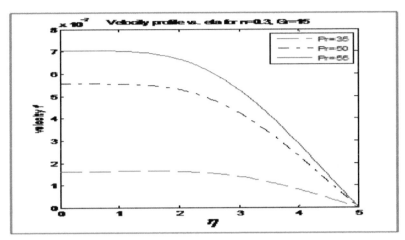

Figure 7.5   Velocity Profile for different values of Pr.

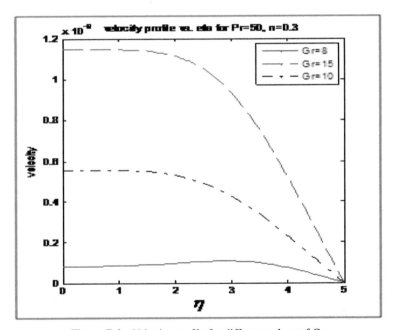

Figure 7.6   Velocity profile for different values of Gr.

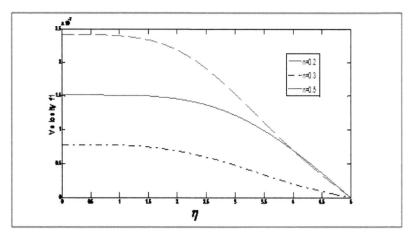

Figure 7.7 Velocity profile for different values of n for Pr=50 & Gr=10.

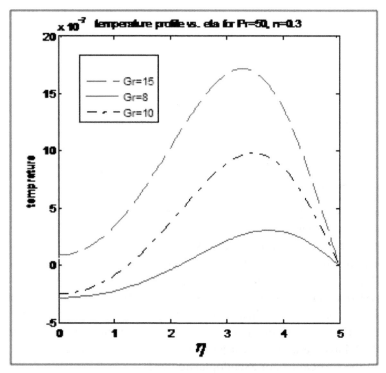

Figure 7.8 Temperature Profile for different values of Gr.

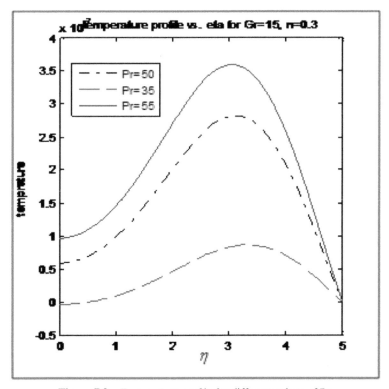

Figure 7.9   Temperature profile for different values of Pr.

7.5 Conclusion

The changes in velocity and temperature profiles are slower for the present jet compared with the Newtonian fluid jet. It is little consistent with the nature of Psudoplasctic fluids. The problem discussed in the present paper is probably applicable to drilling boreholes, natural gas, or oil into the surfaces of the soil using drilling fluids. Also, It may apply to the problems related to a volcanic eruption. Also, a vertical jet flow creates when heartbeats and arteries carry oxygen-rich blood from the heart to all the parts of the human body. In Hawaiian eruptions, lava (magma) immerse from a vent as a fire fountain or lava jet (A huge natural vertical jet of non-Newtonian fluid). It is also an example of the vertical jets of Pseudoplastic fluid.

References

Acrivos, A., Shah, M. J., and Petersen, E. E. 1960 Momentum and heat transfer in laminar boundary-layer flows of non-Newtonian fluids past external bodies, AIChEJ., 6, pp.312-317.

Abd-el-Malek, M. B., Badran, N. A., and Hassan, H. S. 2002 Solution of the Rayleigh problem for a power law non-Newtonian conducting fluid via group method, Int. J. Eng. Sci., 40, pp. 1599–1609.

Adnan, K. A., Hasmani, A. H., Timol, M. G. 2011 A new family of similarity solutions of three dimensional MHD boundary layer flows of non-Newtonian fluids using new systematic grouptheoretic approach, Appl. Math. Sci., 5(27), pp.1325–1336.

Bickley, W. G. 1937 The plane jet, Philos.Mag., 23,pp.727–731.

Conway, William, E. 1967 The two-dimensional vertical jet under gravity, J. Math. Anal. Appl, 19, pp.282–290.

Daniel, N., Riahi 2017 Modeling and computation of nonlinear rotating polymeric jets during Force spinning process, International Journal of Non-Linear Mechanics, 92, pp. 1–7.

Darji, R. M., and Timol, M. G. 2013 Group symmetry and similarity solutions for MHD mixed-convection flow of power-law fluid over a non-linear stretching surface, Int. J. Math. Sci. Comput., 3(2),pp.32–36.

Darji, R. M., and Timol, M. G. 2014 Similarity Analysis for Unsteady Free Convective Boundary Layer Flow of Power Law Fluids via Group Theory, JAMA, 3(1), pp. 9-19.

Decent, S. P., King, A.C., and Wallwork, I. M. 2002 Free jets spun from a prilling tower, J. Eng. Math., 42, pp. 265–282.

Helmholtz, H. L. F. 1868 Über discontinuirliche Flüssigkeitsbewegungen, Monatsbericht Akad. Wiss. Berlin S, pp.215–228.

Jordan, C., Rankin, G. W., and Sridhar, K. 1992 A study of submerged Pseudoplastic laminar jet, Journal of non-Newtonian fluid mechanics, 41(3), pp. 323-337.

Kalathia, N. L. 1975 Laminar jets and flow over a wedge: some problems, A Doctoral Thesis, IIT Kanpur.

Kirchhoff, G. 1869 Zur Theorie freier Flüssigkeitsstrahlen, Z. reine Angew. Math., 70, pp.289-298.

Magan, A. B., Mason, D. P., and Mahomed, F. M. 2016 Analytic solution in parametric form for the two- dimensional free jet of Power-law fluid, International Journal of Non-Linear Mechanics, 85, pp. 94–108.

Moran, M. J., and Gaggioli, R. A. 1968 Similarity analysis via group theory, AIAA J. 6, pp.2014-2016.

Moran, M. J., and Gaggioli, R. A. 1968 A new symmetric formalism for similarity analysis, with application to boundary layer flows, Technical summary report No.918, mathematical Research center, U.S.Army, Madison, Wisconsin.

Moran, M. J., and Gaggioli, R. A. 1968 Reduction of the number of variables in systems of partial differential equations with auxiliary conditions, SIAM J. Appl.Math.16, pp.202-215.

Nita, Jain., and Timol, M. G. 2015 Similarity solutions of quasi three dimensional power law fluids using the method of satisfaction of asymptotic boundary conditions, Alexandria Engineering Journal, 54, pp.725–732.

Parmar, H., and Timol, M. G. 2011 Deductive Group Technique for MHD coupled heat and mass transfer natural convection flow of non-Newtonian power law fluid over a vertical cone through porous medium, Int. J. Appl. Math. Mech., 7 (2), pp.35–50.

Patel, M., and Timol, M. G. 2014 On the Axisymmetrical and Two-Dimensional Jet Flow of an incompressible Pseudo-Plastic Fluids , Int. J. of Appl. Math and Mech., 10(8), pp.45-60.

Patel, M., and Timol, M. G. 2015 Numerical Solution of Two-Dimensional Laminar Circular Jet in a Variable Magnetic Field, International journal of mathematics and scientific computing, 5(1), pp. 14-18.

Patel, M., and Timol, M. G. 2016 Jet with variable fluid properties: Free jet and dissipative jet, International Journal of Non-Linear Mechanics, 85 , pp.54–61.

Patel, M., and Timol, M. G. 2010 The general stress–strain relationship for some different visco-inelastic non-Newtonian fluids, International Journal of Applied Mechanics and Mathematics (IJAMM), 6(12), pp.79-93.

Patel, M., Surati, H., Chanda, M. S., and Timol., M. G. 2013 Models of Various Non-Newtonian Fluids, International E-journal of education and mathematics (IEJEM), 2(4), pp.21-36.

Patel, M., Patel, J., and Timol, M. G. 2017 On the solution of boundary layer flow of Prandtl fluid past a flat surface, Journal of Advanced Mathematics and Applications, l6, pp.1-6.

Patil, V. S., and Timol, M. G. 2012 Three Dimensional Unsteady Incompressible Magneto-Hydrodynamic Boundary Layer Equations of Non-Newtonian Power Law Fluids, JAMA 1(2), pp. 264-268.

Schlichting, H. 1933 LaminareStrahlausbreitung, Z.Angew.Math.Mech., 13, pp.260–263.

Schlichting, H, 1968 Boundary-LayerTheory, McGraw-Hill, NewYork, pp. 218–221.

Schowalter, W. R. 1960 The application of boundary-layer theory to power-law pseudoplastic fluids: similar solutions, AIChEJ, 6, pp.24-28.

Stehr, H., and Schneider, W. 2000 Jet flows in non-Newtonian fluids, Z. angew. Math. Phys., 51, pp. 922-941.

Wallwork, I. M., Decent, S.P., King, A. C., and Schulkes, R.M.S.M. 2002 The trajectory and stability of a spiraling liquid jet: Part I. inviscid theory, J. Fluid Mech., 459, pp. 43–65.

Wilkinson, W. L. 1960 Non-Newtonian fluids: Fluid mechanics, mixing and heat transfer, Pergamon Press.

8

Analytical Solutions For Hydromagnetic Flow of Chemically Reacting Williamson Fluid Over a Vertical Cone and Wedge with Heat Source/Sink

A. Subramanyam Reddy[1*], S. Srinivas[2], and Anant Kant Shukla[3]

[1]Department of Mathematics, School of Advanced Sciences, VIT, Vellore-632014, Tamil Nadu, India
[2]Department of Mathematics, School of Sciences and Languages, VIT-AP University, Amaravati - 522 237, India
[3]Department of Mathematics, School of Advanced Sciences and Languages, VIT Bhopal University, Sehore – 466114, India
E-mail: anala.subramanyamreddy@gmail.com
*Corresponding Author

Abstract

This paper presents the analysis of the hydromagnetic flow of chemically reacting Williamson fluid over a vertical cone and wedge with a heat generation/absorption. Suitable similarity transformations are invoked to convert the governing equations of the present model into ordinary differential equations. The homotopy analysis method (HAM) with a non-homogeneous term is utilized to get the analytical solutions through the minimization of the average square residual error. The obtained analytical solutions are in good agreement with the results available in the literature.The analysis depicts that the velocity profiles decrease with a rise in Williamson parameter and applied magnetic field. An increase in applied magnetic field raises the temperature while there is a fall in concentration with an enhancement in chemical reaction parameters.

155

Keywords: Williamson fluid, Cone/wedge, Hartmann number, Chemical reaction, HAM with the non-homogeneous term.

8.1 Introduction

The work of heat transfer in rheological fluids has gotten considerable attention from researchers because of its applications in the biological, chemical, pharmaceutical, and food processing industries [1–3]. Cheng [4] analyzed the boundary layer flow of power-law fluid with yield stress over a vertical plate. Benazir et al. [5] have made a comparison of Casson fluid flow with heat mass transfer between vertical cone and flat plate by employing a finite difference method. Nadeem and Hussain [6] employed optimal HAM to get the series solutions for Williamson fluid over the exponential sheet. Dapra and Scarpi [7] examined the pulsating flow of Williamson fluid in a rock fracture. The problem of MHD flow of a Williamson fluid over a stretching sheet with Cattanneo-Christo heat flux model was explored by Salahuddin et al. [8].

The studies related to the chemical reaction with heat and mass transfer have a great impact (applications) in science, engineering and industries. For example, chemical engineering, metallurgy, combustion systems, nuclear reactor safety, solar collectors. The industries are designed to find a high-value product through the cheaper raw material with the help of chemical reaction [9–12]. Malik et al. [13] employed the Keller box method to address the impacts of homogeneous-heterogeneous reactions on Williamson fluid flow through a stretching cylinder. Rashidi et al. [14] obtained Lie group solutions for free convective chemically reacting nanofluid flow past a horizontal plate with a porous medium. Srinivas et al. [15] demonstrated the viscous fluid flow through a porous pipe with expanding or contracting walls with chemical reaction by using HAM. The impact of chemically reacting rheological fluid flow over a vertical cone and the flat plate was numerically examined by Mythili and Sivaraj [16].

The analysis of boundary layer flow with heat transfer through a rotating cone arises in several applications of engineering and chemical industry like rotating heat exchangers, nuclear reactor cooling systems, geothermal reservoirs, design of canisters for nuclear waste disposal, and so on [17–18]. Ramanaiah and Kumaran [19] first explored the Darcy-Brinkman free convection about a cone and wedge in a porous medium. The analysis of the convection flow over a vertical cone/wedge with a heat source was done by Vajravelu and Nayfeh [20]. Some more studies have been carried out related to the fluid flow over a cone and wedge ([21–24] several references therein).

The above literature shows that no study regarding chemically reacting MHD flow of Williamson fluid over a vertical cone/wedge has been examined so far. Because of the importance of this kind of analysis in industry and engineering, here an effort is made to analyze the impacts of chemical reaction and heat source/sink on the MHD flow of Williamson fluid over a vertical cone and wedge by employing HAM with a non-homogeneous term. In recent years, HAM is being used widely as a tool to develop approximate analytical solutions for solving nonlinear problems [25–30]. Series solutions of the present work are obtained by following [31, 32]. Convergence parameters present in the HAM with a non-homogeneous technique are obtained by minimizing the average square residual error. The impact of various parameters has been discussed through graphical and tabular results.

8.2 Formulation of the Problem

We consider Williamson fluid which is electrically conducted over a vertical cone and wedge. As presented in Figure 8.1, the center (origin) of the Cartesian coordinate system is placed at the vertex of the full cone, whereas the horizontal line (x-axis) is along the surface and the vertical line is (y-axis)

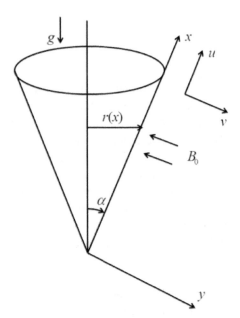

Figure 8.1   The coordinate system of the model.

is orthogonal to the surface. Based on the above settings, and with the following assumptions the governing equations are [2, 3, 6]

- Flow is laminar and incompressible.
- An applied magnetic field of strength B_0 is considered normal to the surfaces.
- Viscous and magnetic dissipations are neglected.
- Heat generation/absorption and Chemical reaction are considered.
- Boussinesq approximations are considered.

$$\frac{\partial}{\partial x}(r^n u) + \frac{\partial}{\partial y}(r^n v) = 0 \tag{8.5}$$

$$u\frac{\partial u}{\partial x} + v\frac{\partial u}{\partial y} = v\frac{\partial^2 u}{\partial y^2} + \sqrt{2}v\Gamma\frac{\partial u}{\partial y}\frac{\partial^2 u}{\partial y^2} - \frac{\sigma B_0^2}{\rho}u$$
$$+ (g\beta_T T - g\beta_T T_\infty)\cos\alpha + (g\beta_C C - g\beta_C C_\infty)\cos\alpha \tag{8.6}$$

$$u\frac{\partial T}{\partial x} + v\frac{\partial T}{\partial y} = \frac{\kappa}{\rho C_p}\frac{\partial^2 T}{\partial y^2} + \frac{Q_0}{\rho C_p}(T - T_\infty) \tag{8.7}$$

$$u\frac{\partial C}{\partial x} + v\frac{\partial C}{\partial y} = D\frac{\partial^2 C}{\partial y^2} - k_1(C - C_\infty). \tag{8.8}$$

Here the velocity components u, v are in the of direction x, y respectively, ρ, v are density, kinematic viscosity, respectively, β_T thermal expansion coefficient, β_C is concentration expansion coefficient, σ is electrical conductivity, α is the half angle of cone or wedge, κ is thermal conductivity, Q_0 is heat generation/absorption coefficient, g is acceleration due to gravity, C_p is specific heat at constant pressure, T, C are temperature and concentration of the fluid, r is local radius of the cone, T_∞, C_∞ are temperature and concentration at infinity, D is coefficient of mass diffusivity and k_1 is the rate of first order chemical reaction ($k_1 > 0$ for destructive chemical reaction and $k_1 < 0$ is generative chemical reaction). $n = 1$ and $n = 0$ represent to flow over a vertical cone and flow over a wedge respectively,

The appropriate boundary conditions are

$$u(x, 0) = 0, \quad v(x, 0) = 0, \quad T(x, 0) = T_w = T_\infty + Ax^{m_1},$$
$$C(x, 0) = C_w = C_\infty + Bx^{m_2} \tag{8.9}$$

$$u\left(x, \infty\right) = 0, \quad v\left(x, \infty\right) = 0, \quad T\left(x, \infty\right) = 0, C\left(x, \infty\right) = 0 \quad (8.10)$$

where T_w, C_w are surface temperature and concentration, A, B are known constants and m_1, m_2 are surface temperature and concentration parameters respectively. It is worth mentioning that $m_1 = m_2 = 0$ represents the constant wall temperature and concentrations respectively.

Now, by following the non-dimensional variables [20–21]

$$\eta = \frac{y}{L}, \ u = \frac{vxf'\left(\eta\right)}{L^2}, \ v = \frac{-f\left(\eta\right)\left(n+1\right)v}{L},$$
$$T = T_\infty + \theta\left(\eta\right)\left(T_w - T_\infty\right), \ C = C_\infty + \phi\left(\eta\right)\left(C_w - C_\infty\right) \quad (8.11)$$

Equations (8.6)–(8.10) are reduces to

$$f''' + \left(n+1\right)ff'' + \gamma f''f''' - f'^2 - M^2 f' + Gr_x\theta + Gc_x\phi = 0 \quad (8.12)$$

$$\theta'' + \left(n+1\right)\Pr f\theta' - m_1 \Pr f'\theta + Q\theta = 0 \quad (8.13)$$

$$\phi'' + \left(n+1\right)Scf\phi' - m_2 Scf'\phi - K\phi = 0 \quad (8.14)$$

$$f = 0, f' = 0, \theta = 1, \phi = 1 at \eta = 0 \quad (8.15)$$
$$f = 0, f' = 0, \theta = 0, \phi = 0 as \eta \to \infty \quad (8.16)$$

where $\gamma = \frac{\sqrt{2}\Gamma xv}{L^3}$ is local Williamson parameter, $M = \frac{\sqrt{\sigma}B_0 L}{\sqrt{\mu}}$ is Hartmann number, $Gr_x = \frac{L^4 g\beta(T_w - T_\infty)\cos\alpha}{v^2 x}$ is local thermal Grashof number, $Gc_x = \frac{L^4 g\beta_C(C_w - C_\infty)\cos\alpha}{v^2 x}$ is local mass Grashof number, $\Pr = \frac{\mu C_p}{\kappa}$ is Prandtl number, $Q = \frac{Q_0 L^2}{\kappa}$ is heat source/sink parameter (for heat generation $Q > 0$ and for heat absorption $Q < 0$), $Sc = \frac{v}{D}$ is the Schmidt number and $K = \frac{k_1 L^2}{v}$ is chemical reaction parameter, L is characteristic length and primes represent the derivatives with respect to η.

Further, the dimensionless Nusselt number (Nu) and Sherwood number (Sh) are given as

$$Nu = -\theta'\left(0\right), Sh = -\phi'\left(0\right). \quad (8.17)$$

8.3 Solution of the Problem

In the current section, analytical solutions for the coupled system (8.12)-(8.14) with the corresponding B.Cs (8.15)-(8.16) are obtained with the help of HAM with a non-homogeneous term. For approximate analytical solutions,

Table 8.1   Comparison of present numerical values of $f''(0)$ and $\theta'(0)$ with Chamkha [21] and Vajravelu and Nayfeh [20] for $n = 1, \gamma = 0, \Pr = 0.3, Q = -5, Gr_x = -0.5, M = 1, Gc_x = K = Sc = m_2 = 0.$

m_1	$f''(0)$			$\theta'(0)$		
	[21]	[20]	Current	[21]	[20]	Current
−2.1	−0.155592	−0.155592	−0.155541	−2.238739	−2.237475	−2.23792
2.1	−0.155995	−0.156001	−0.156057	−2.234070	−2.232780	−2.23246

f_0, θ_0, ϕ_0 are initial guesses and L_1, L_2, L_3 are auxiliary linear operators respectively which are given as

$$f_0(\eta) = 0, \quad \theta_0(\eta) = e^{-\eta}, \quad \phi_0(\eta) = e^{-\eta} \tag{8.18}$$

$$L_1(f) = \frac{d^3 f}{d\eta^3} - \frac{df}{d\eta}, \quad L_2(\theta) = \frac{d^2 \theta}{d\eta^2} - \theta, \quad L_3(\phi) = \frac{d^2 \phi}{d\eta^2} - \phi. \tag{8.19}$$

For simplicity, the non-homogeneous term is taken in the form of $a_1 e^{-\eta}$. The selection of the non-homogeneous term has been discussed by Shukla et al. [28–29]. The convergence control parameters h, a_1 are chosen by minimizing the average square residual error. For more details on HAM with a non-homogeneous term, the reader can refer to [28–29]. Further, to check the accuracy of obtained results, we have compared the results of [20] and [21] with the present results for Newtonian fluid case (i.e. $\gamma = 0$), and are presented in Table 8.1. The comparison reveals that there is a good agreement between the results.

8.4 Results and Discussion

To examine the impact of parameters that are involved in the present model on dimensionless flow variables, the results are presented in Figures 8.2–8.4. Further, the numerical values for dimensionless Nusselt and Sherwood numbers are tabulated in Table 8.2. In these computations, we keep the values as $\gamma = 2$, $M = 2$, $Gr_x = Gc_x = 2$, $\Pr = 0.71$, $Q = -1, m_1 = m_2 = 1$, $Sc = 0.78$, $K = 1$. The variations in $f'(\eta)$ (velocity) for a set value of $\gamma, M, Q\ K$ are depicted in Figure 8.2a- 8.2d. Figure 8.2a reveals that $f'(\eta)$ decreases for a given increase in γ for cone as well as a wedge. It is also noticed that the thickness of the momentum boundary layer is higher for Newtonian fluid ($\gamma = 0$) in comparison to non-Newtonian fluid ($\gamma \neq 0$). Figure 8.2b is revealed that there is a fall in velocity profiles with a rise in M. The reason for the fall in the velocity is because the retarding forces upon the application of the magnetic field acted opposite to the direction of the flow.

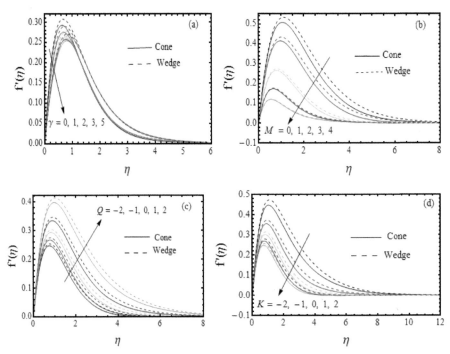

Figure 8.2   Velocity profiles (a) impact of γ (b) impact of M (c) impact of Q (d) impact of K.

Figure 8.2c elucidates that the velocity profiles increase when a heat source $(Q > 0)$ is increased while there is a drop in velocity with enhancing heat sink $(Q < 0)$. Figure 8.2d shows that there is a rise in velocity profiles for the case of generative chemical reaction $(K < 0)$ while there is a fall for the case of a destructive chemical reaction $(K > 0)$. Further, from Figure 8.2 one can say that the velocity is more prominent (higher) in wedge as compared to a cone.

The effects of Q, M, m_1, Gr_x on temperature profiles $\theta(\eta)$ are presented in Figure 8.3a- 8.3d. Figure 8.3 elucidates in the both cases of cone and wedge, $\theta(\eta)$ enhances with enhancing heat generation, whereas enhancing heat absorption reduces $\theta(\eta)$. From Figure 8.3b one can notice that a rise in M accelerates the temperature. It is due to the fact that a rise in applied magnetic field (Lorentz forces) increases the thermal boundary layer thickness. Figure 8.3c presents the effect of surface temperature parameter m_1 on $\theta(\eta)$. Its depicted that an increase in m_1 decelerates $\theta(\eta)$. Figure 8.3d demonstrated the impact of Gr_x on $\theta(\eta)$. One can infer that for the both

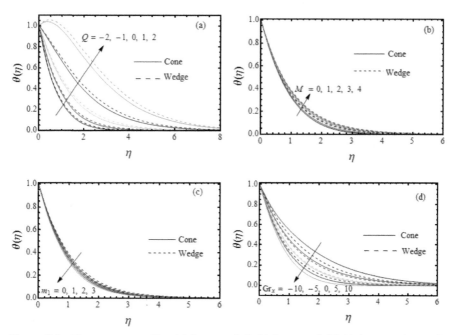

Figure 8.3   Temperature profiles (a) impact of Q (b) impact of M (c) impact of m_1 (d) impact of Gr_x.

cases of cone and wedge, enhancing Gr_x drops the temperature profiles for $Gr_x > 0$ while it enhances temperature profiles for $Gr_x < 0$.

The influence of chemical reaction parameter (K), surface concentration parameter (m_2), and local mass Grashof number (Gc_x) on the concentration profiles $\phi(\eta)$ is presented in Figure 8.4 for the cases of cone and wedge. Figure 8.4a reveals that a rise in destructive chemical reactions reduces the concentration. This is because an increase in destructive reaction reduces the concentration boundary layer thickness which decreases the concentration of diffusing species. From the same figure, the opposite behavior is detected in the case of the generative chemical reaction. From Figure 8.4b one can infer that there is a decrease in concentration profiles with an increase in m_1. From Figure 8.4c, a similar behavior is noticed with an enhancement of Gc_x. Table 8.2 analyses the impacts of γ, M, K, Q, Gr_x, Gc_x on Nusselt number and Sherwood number distributions. It infers that for both the cases of cone and wedge, there is a fall in Nu with a rise in γ, M, K and heat generation whereas it rises with an enhancement in Gr_x, Gc_x and heat absorption. Further, a rise in K, Gr_x, Gc_x and heat generation accelerates Sh whereas γ, M and heat absorption decelerates Sh.

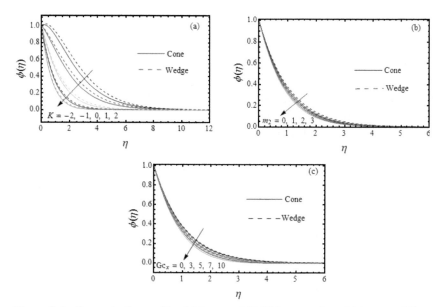

Figure 8.4 Concentration profiles (a) impact of K (b) impact of m_2 (c) impact of Gc_x.

Table 8.2 Effects of γ, M, K, Q, Gr_x, Gc_x on Nu and Sh.

Physical parameters	Values	Nu		Sh	
		Cone	Wedge	Cone	Wedge
γ	0	0.99425	0.96294	1.03715	1.00477
	1	0.98445	0.95526	1.02703	0.99683
	2	0.97658	0.94906	1.01888	0.99040
M	0	1.01165	0.98410	1.03366	1.00566
	2	0.97415	0.95534	0.99531	0.97617
	4	0.93761	0.92741	0.95790	0.94751
K	0	0.98214	0.95858	0.65107	0.62031
	1	0.97539	0.95335	1.00286	0.98039
	2	0.97025	0.94938	1.28320	1.26609
Q	−2	1.22833	1.21337	0.99156	0.97328
	−1	0.97415	0.95534	0.99531	0.97617
	0	0.66495	0.64067	1.00000	0.97976
	1	0.27217	0.24008	1.00591	0.98429
	2	−0.24951	−0.29278	1.01343	0.99004
Gr_x	0	0.93730	0.92719	0.95752	0.94724
	5	1.02304	0.99337	1.04549	1.01529
	10	1.08937	1.04672	1.11367	1.07021
Gc_x	0	0.95365	0.94787	0.95846	0.95261
	3	0.98714	0.97319	0.99270	0.97855
	5	1.00829	0.98930	1.01433	0.99508

8.5 Conclusion

In the present work, we have investigated the behavior of chemically reacting MHD flow of Williamson fluid over a vertical cone and wedge in the presence of a heat source/sink. HAM with a non-homogeneous term is utilized to get analytical solutions for flow variables. The results of [20–21] are verified in this investigation in the case of a Newtonian fluid. The analysis highlights that the velocity is reduced with higher values of Williamson parameter and applied magnetic field whereas a rise in heat source enhances the velocity. The temperature rises with higher values of Hartmann number and heat source while the temperature falls with an enhancement of heat sink and surface temperature parameter. Further, the concentration decreases for a given increase in a destructive chemical reaction, surface concentration parameter, and local mass Grashof number.

References

[1] Nadeen, S. Saleem, Series solution of unsteady Eyring Powel nanofluid flow on a rotating cone Indian J. Pure Appl. Phys. 52 (2014) 725-737.

[2] R. S. R. Gorla, J. Gireesha, Dual solutions for stagnation-point flow and convective heat transfer of a Williamson nanofluid flow past a stretching/shrinking sheet, Heat Mass Transfer 52 (2016) 1153-1162.

[3] S. Nadeem, S. T. Hussain, G. Lee, Flow of a Williamson fluid over a stretching sheet, Brazilian J. Chem. Eng. 30 (2013) 619-625.

[4] C-Y. Cheng, Soret and Dufour effects on free convection boundary layers of non-Newtonian power law fluids with yield stress in porous media over a vertical plate variable wall heat and mass fluxes, Int. Commun. Heat Mass transfer 38 (2011) 615-619.

[5] A.J. Benazir, R. Sivaraj, M.M. Rashidi, Comparison between Casson fluid flow in the presence of heat and mass transfer from a vertical cone and flat plate, J. Heat Transfer 138 (2016) 112005-7.

[6] S. Nadeem, S.T. Hussain, Heat transfer analysis of Williamson fluid over exponentially stretching surface, Appl. Math. Mech. 35 (2014) 489-502.

[7] I. Dapra, and G. Scarpi, Perturbation solution for pulsatile flow of a non-Newtonian Williamson fluid in a rock fracture, Int. J. Rock Mech. Mining Sci. 44 (2007) 271-278.

[8] T. Salahuddin, M.Y. Malik, A. Hussain, S. Bilal, M. Awais, MHD flow of Cattanneo-Christov heat flux model for Williamson fluid over

a stretching sheet with variable thickness: Using numerical approach, J. Magn. Magn. Mater. 401 (2016) 991-997.

[9] T. Hayat, Z. Abbas, Channel flow of a Maxwell fluid with chemical reaction, Zeitschrift für angewandte Mathematik und Physik 59 (2008) 124-144.

[10] K. Vajravelu, K.V. Prasad, N.S. P. Rao, Diffusion of a chemically reactive species of a power-law fluid past a stretching surface, Comput. Math. Appl. 62 (2011) 93-108.

[11] A. Subramanyam Reddy, S. Srinivas, T.R. Ramamohan, Analysis of heat and chemical reaction on an asymmetric laminar flow between slowly expanding or contracting walls, Heat Transfer-Asian research 42 (2013) 422-443.

[12] K. U. Rehman, A. A. Khan, M. Y. Malik, U. Ali, M. Naseer, Numerical analysis subject to double stratification and chemically reactive species on Williamson dual convection fluid flow yield by an inclined stretching cylindrical surface, Chin. J. Phys. (2017) doi: 10.1016/j.cjph.2017.05.003.

[13] M. Y. Malik, T. Salahuddin, A. Hussain, S. Bilal, M. Awais, Homogeneous-heterogeneous reactions in Williamson fluid model over a stretching cylinder by using Keller box method, AIP Adv. 5 (2015) 107227-14.

[14] M. M. Rashidi, E. Momoniat, M. Ferdows, A. Basiriparsa, Lie group solution for free convective flow of a nanofluid past a chemically reacting horizontal plate in a porous media, Math. Probl. Eng. 2014 (2014), 239082, doi.org/10.1155/2014/239082.

[15] S. Srinivas, A. Subramanyam Reddy, T. R. Ramamohan, Mass transfer effects on viscous flow in an expanding or contracting porous pipe with chemical reaction, Heat transfer-Asian Research 44 (2015) 552-567.

[16] D. Mythili, R. Sivaraj, Influence of higher order chemical reaction and non-uniform heat source/sink on Casson fluid flow over a vertical cone and flat plate, J. Mol. Liq. 216 (2016) 466-475.

[17] S. Nadeem, S. Saleem, Analytical treatment of unsteady mixed convection MHD flow on a rotating cone in a rotating frame, J. Taiwan Inst. Chem. Eng. 44 (2013) 596-604.

[18] S. Nadeem, S. Saleem, Analytical study of third grade fluid over a rotating vertical cone in the presence of nanoparticles, Int. J. Heat Mass Transfer 85 (2015) 1041-1048.

[19] G. Ramanaiah, V. Kumaran, Darcy-Brinkman free convection about a wedge and a cone subjected to a mixed thermal boundary condition, Int. J. Math. Math. Sci. 15 (1992) 789-794.

[20] K. Vajravelu, J. Nayfeh, Hydromagnetic convection at a cone and a wedge, Int. Commun. Heat Mass Transfer 19 (1992) 701-710.

[21] A. J. Chamkha, Non-Darcy hydromagnetic free convection from a cone and a wedge in porous media, Int. Commun. Heat Mass Transfer 23 (1996) 875-887.

[22] S. M. Al-Harbi, Numerical study of natural convection heat transfer with variable viscosity and thermal radiation from a cone and wedge in porous media, Appl. Math. Comput. 170 (2005) 64-75.

[23] A. Hassanien, A. S. Essawy, N. M. Moursy, Variable viscosity and thermal conductivity effects on heat transfer by natural convection from a cone and a wedge in porous media Archives Mech. 55 (2003) 345-355.

[24] S. Srinivas, A. Subramanyam Reddy, and A. Vijayalakshmi, Flow and heat transfer of Williamson fluid over a cone and wedge with magnetic field, AIP Conference Proceedings 1830, 020002 (2017); doi: 10.1063/1.4980865.

[25] S. J. Liao, Homotopy analysis method in nonlinear differential equations, Springer & Higher Education Press, Heidelberg, (2012).

[26] Z. Liu, S. J. Liao, Steady-state resonance of multiple wave interactions in deep water, J. Fluid Mech. 742 (2014) 664-700.

[27] T. Muhammad, T. Hayat, A. Alsaedi, A. Qayyum, Hydromagnetic unsteady squeezing flow of Jeffery fluid between two parallel plates, Chin. J. Phys. (2017) doi: 10.1016/j.cjph.2017.05.008.

[28] A. K. Shukla, T. R. Ramamohan, S. Srinivas, Homotopy analysis method with a non-homogeneous term in the auxiliary linear operator, Commun. Nonlinear Sci. Numerical Simulat. 17 (2012) 3776-3787.

[29] S. Srinivas, A. K. Shukla, T. R. Ramamohan, A. S. Reddy, Influence of thermal radiation on unsteady flow over an expanding or contracting cylinder with thermal-diffusion and diffusion-thermo effects, J. Aerospace Eng. 28 (2014) 04014134-10.

9

Aboodh Transform Homotopy Perturbation Method for Solving Newell–Whitehead–Segel Equation

Haresh P. Jani and Twinkle R. Singh

Applied Mathematics and Humanities Department, Sardar Vallabhbhai National Institute of Technology, Surat-395007 (Gujarat), India
E-mail: hareshjani67@gmail.com; twinklesingh.svnit@gmail.com

Abstruct

This work deals with the Aboodh transform with the homotopy perturbation method (HPM). HPM is presented to solve nonlinear differential equations. Aboodh transform is restricted to linear equations, and HPM is an efficient and dominant technique for nonlinear differential equations to find approximate solutions. NWSE is an essential model that arises in fluid mechanics. The final outcomes execute the excellent agreement with the exact solution. Graphic descriptions of obtained results are also presented.

Keywords : Nonlinear differential equation, Aboodh transform, Homotopy perturbation method, Newell–whitehead–segel equation (NWSE).

9.1 Introduction

Nowadays, there are many developments to solve nonlinear differential equations. Many nonlinear differential equations do not have an analytic solution [1–5]. In this work, ATHPM is standing to obtain the solution of NWSE. NWSE epitomizes many stripe patterns, like ripples in sand in 2 dimensional systems, and stripes of seashells. NWSE is written as, [6–12]

167

$$\frac{\partial U(\theta,\tau)}{\partial \tau}=w\frac{\partial^2 U(\theta,\tau)}{\partial \theta^2}+aU(\theta,\tau)-bU(\theta,\tau)^r \qquad (9.1)$$

where real numbers w,b and a > 0 , $U(\theta,\tau)$ is the fluid flow velocity in an infinite long conduit with a tiny diameter. $\frac{\partial U(\theta,\tau)}{\partial \tau}$ describes the changes of U with time at a fixed position, $\frac{\partial^2 U(\theta,\tau)}{\partial \theta^2}$ expresses the changes of $U(\theta,\tau)$ with θ at fixed time τ and $aU-bU^r$ is releasing parameter as a source term. Solution of NWSE has been calculated by Variational iteration method [7, 12, 13], Homotopy perturbation method [9, 11], A domain decomposition method [14], Mohand transform [16] and Homotopy analysis method [10]. The NWSE has massive applicability in mechanical engineering and biology. Aboodh transform was founded by Khalid Aboodh to solve some differential equations in the time field [1, 2]. NWSE can be described Rayleigh-Benard convection and Zeldovich equation in some cases while taking different values of a, b & r. Biologically, Rayleigh-Benard convection happens in a planar parallel layer of fluid heated from the inferior side. The fluid forms a natural model of convection cells recognized as Benard cells.

This work combines two powerful hybrid techniques, the Aboodh transform, and HPM, to determine approximate analytic solutions of nonlinear DE. The primary purpose of this work is to extend the application of ATHPM. The direct benefit of this hybrid approach is its combination and ability to achieve the solution of nonlinear differential equations. The study results obtain excellently and match the exact solution.

9.2 Basic Definition of Aboodh Transform

We investigate functions in the set A given by the Aboodh transform, which is defined for functions of exponential order [1, 2].

$$A=\{f(\tau):\exists M, k_1, k_2> 0, |f(\tau)|<C\,e^{-v}\} \qquad (9.2)$$

The constant C must be a finite integer for a given function in the set A, and k_1, k_2 may be infinite or finite.

The Aboodh transform denoted by operator A(.) is defined by the integral equation

$$A[f(\tau)](v)=K(v)=\frac{1}{v}\int_0^\infty f(\tau)\,e^{-v\tau}d\tau,\ \tau{\geq}0, k_1{\leq}v{\leq}k_2 \qquad (9.3)$$

In the argument of the function f, the variable v is utilized to factor the variable tau. For the Aboodh transform to exist, the function must be piecewise continuous and of exponential order.

9.2.1 Some Properies of Aboodh Transform

1. $A[1] = \frac{1}{v^2}$
2. $A[\tau] = \frac{1}{v^3}$
3. $A[\tau^n] = \frac{n!}{v^{n+2}}$
4. $A[e^{a\tau}] = \frac{1}{v^2 - av}$
5. $A[sina\tau] = \frac{a}{v(v^2 + a^2)}$
6. $A[cosa\tau] = \frac{1}{v^2 + a^2}$
   if $A[f(\tau)] = K(v)$ then
7. $A[f^{(n)}(\tau)] = v^{(n)} K(v) - \prod_{k=0}^{n-1} \frac{f^{(k)}(0)}{v^{2-n+k}}$

Definition 1 Inverse Aboodh transform of a function $f(\tau)$, if $A[f(\tau)] = K(v)$ then $f(\tau) = A^{-1}[K(v)]$ where $\tau \in (0, \infty)$

9.3 Idea of Aboodh Transform Homotopy Perturbation Method

NWSE model is

$$\frac{\partial U}{\partial \tau} = w \frac{\partial^2 U}{\partial \theta^2} + aU - bU^r \qquad (9.4)$$

Where the real numbers w,b and a > 0 and $r \in \mathbb{N}$ with $U(0, \tau) = c(\theta)$
Apply Aboodh transform in above equation, [6]

$$A[\frac{\partial U}{\partial \tau}] = A[w \frac{\partial^2 U}{\partial \theta^2} + aU - bU^r]$$

By Aboodh transform,

$$v\{K(v) - \frac{1}{v^2} U(0,0)\} = A[w \frac{\partial^2 U}{\partial \theta^2} + aU - bU^r]$$

which can be written as

$$K(v) = \frac{1}{v^2} U(0,0) + \frac{1}{v} A[w \frac{\partial^2 U}{\partial \theta^2} + aU - bU^r]$$

Now, by inverse Aboodh transform

$$u(\theta, \tau) = U(\theta, 0) + A^{-1}[\frac{1}{v}A(w\frac{\partial^2 U}{\partial\theta^2} + aU - bU^r)]$$

Assuming the solution is of the form

$$U = U_0 + pU_1 + p^2 U_2 + \dots$$

To consider the nonlinear operator, we apply HPM [3, 4, 5]

$$\sum_{n=0}^{\infty} p^n U_n(\theta, \tau) = U(\theta, 0) + pA^{-1}[\frac{1}{v}A(R\sum_{n=0}^{\infty} p^n H_n(\theta, \tau)$$

$$- N\sum_{n=0}^{\infty} p^n H_n(\theta, \tau))]$$

equating coefficients of powers of p on both sides, we get

$$p^0 : U_0(\theta, \tau) = U(\theta, 0)$$

$$p^1 : U_1(\theta, \tau) = -A^{-1}[\frac{1}{v}A(RU_0(\theta, \tau) - NU_0(\theta, \tau))]$$

$$p^2 : U_2(\theta, \tau) = -A^{-1}[\frac{1}{v}A(RU_1(\theta, \tau) - NU_1(\theta, \tau))]$$

$$p^3 : U_3(\theta, \tau) = -A^{-1}[\frac{1}{v}A(RU_2(\theta, \tau) - NU_2(\theta, \tau))]$$

$$p^4 : U_4(\theta, \tau) = -A^{-1}[\frac{1}{v}A(RU_3(\theta, \tau) - NU_3(\theta, \tau))].$$

Approximate solution is,

$$U(\theta, \tau) = \sum_{i=0}^{i=\infty} U_i$$

9.4 Some Illustrations

Example 9.1 In equation (4), for a=-3, w=1 and b=0 the NWSE is written as
:

$$\frac{\partial U}{\partial \tau} = \frac{\partial^2 U}{\partial \theta^2} - 3U$$

with initial condition $u(\theta, 0) = e^{2\theta}$ [8]

Apply Aboodh transform

$$A[\frac{\partial U}{\partial \tau}] = A[\frac{\partial^2 U}{\partial \theta^2} - 3U]$$

$$k(\theta, \tau) = \frac{1}{v^2}U(\theta, 0) + \frac{1}{v}A[\frac{\partial^2 U}{\partial \theta^2} - 3U]$$

Apply inverse Aboodh transform

$$U(\theta, \tau) = U(\theta, 0) + A^{-1}[\frac{1}{v}A(\frac{\partial^2 U}{\partial \theta^2} - 3U)]$$

By HPM

$$U_0 = U(\theta, 0) = e^{2\theta}$$

$$p^1:U_1 = A^{-1}[\frac{1}{v}A(\frac{\partial^2 U_0}{\partial \theta^2} - 3U_0)] = \frac{e^{2\theta}\tau}{1!}.$$

$$p^2:U_2 = A^{-1}[\frac{1}{v}A(\frac{\partial^2 U_1}{\partial \theta^2} - 3U_1)] = \frac{e^{2\theta}\tau^2}{(2)!}.$$

$$p^3:U_3 = A^{-1}[\frac{1}{v}A(\frac{\partial^2 U_2}{\partial \theta^2} - 3U_2)] = \frac{e^{2\theta}\tau^3}{(3)!}.$$

Approximate solution is,

$$U(\theta, \tau) = \sum_{i=0}^{i=\infty} U_i$$

$$U(\theta, \tau) = e^{\theta} + \frac{e^{2\theta}\tau^1}{1!} + \frac{e^{2\theta}\tau^2}{(2)!} + \frac{e^{2\theta}\tau^3}{(3)!} + ...$$

The classical NWSE has the solution

$$U(\theta, \tau) = e^{2\theta + \tau}$$

Which is a akin solution to the laplace Adomian decomposition method (LADM) [8].

Example 9.2 In equation (4), for a=-2, w=1 and b=0 the NWSE is written as :

$$\frac{\partial U}{\partial \tau} = \frac{\partial^2 U}{\partial \theta^2} - 2U$$

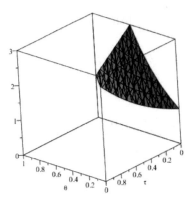

Figure 9.1   Approximate solution for E.g.1

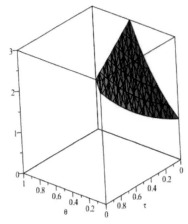

Figure 9.2   Approximate solution for E.g.1

with initial condition $u(\theta, 0) = e^{\theta}$ [15]

   Apply Aboodh transform

$$A[\frac{\partial U}{\partial \tau}] = A[\frac{\partial^2 U}{\partial \theta^2} - 2U]$$

$$k(\theta, \tau) = \frac{1}{v^2} U(\theta, 0) + \frac{1}{v} A[\frac{\partial^2 U}{\partial \theta^2} - 2U]$$

Apply inverse Aboodh transform

$$U(\theta, \tau) = U(\theta, 0) + A^{-1}[\frac{1}{v} A(\frac{\partial^2 U}{\partial \theta^2} - 2U)]$$

By HPM

$$U_0 = U(\theta, 0) = e^\theta$$

$$p^1: U_1 = A^{-1}[\frac{1}{v}A(\frac{\partial^2 U_0}{\partial \theta^2} - 2U_0)] = \frac{-e^\theta \tau}{1!}$$

$$p^2: U_2 = A^{-1}[\frac{1}{v}A(\frac{\partial^2 U_1}{\partial \theta^2} - 2U_1)] = \frac{e^\theta \tau^2}{(2)!}$$

$$p^3: U_3 = A^{-1}[\frac{1}{v}A(\frac{\partial^2 U_2}{\partial \theta^2} - 2U_2)] = \frac{-e^\theta \tau^3}{(3)!}$$

Approximate solution is,

$$U(\theta, \tau) = \sum_{i=0}^{i=\infty} U_i$$

$$U(\theta, \tau) = e^\theta + \frac{-e^\theta \tau^1}{1!} + \frac{e^\theta \tau^2}{(2)!} + \frac{-e^\theta \tau^3}{(3)!} + ...$$

The classical NWSE has the solution

$$U(\theta, \tau) = e^{\theta - \tau}$$

Which is a akin solution to the Adomian decomposition method (ADM) [15].

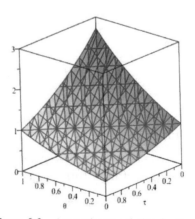

Figure 9.3 Approximate solution for E.g.1

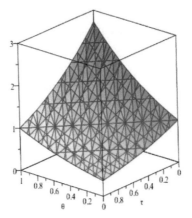

Figure 9.4   Exact solution for E.g. 2

Example 9.3 In equation (4), for a=1, w=1, r=2 and b=1 NWSE is written as :

$$\frac{\partial U}{\partial \tau} = \frac{\partial^2 U}{\partial \theta^2} + U - U^2$$

with initial condition $U(\theta, 0) = \dfrac{1}{(1 + e^{\frac{\theta}{\sqrt{6}}})^2}$ [15]

Apply Aboodh transform

$$A[\frac{\partial U}{\partial \tau}] = A[\frac{\partial^2 U}{\partial \theta^2} + U - U^2]$$

$$k(\theta, \tau) = \frac{1}{v^2} U(\theta, 0) + \frac{1}{v} A[\frac{\partial^2 U}{\partial \theta^2} + U - U^2]$$

Apply inverse Aboodh transform

$$U(\theta, \tau) = U(\theta, 0) + A^{-1}[\frac{1}{v} A(\frac{\partial^2 U}{\partial \theta^2} + U - U^2)]$$

By HPM, costruct the homotopy and let $U = U_0 + pU_1 + p^2 U_2 + \dots$ is solution and we take p→1 for final solution

$$p^1 : \frac{\partial U_1}{\partial \tau} + \frac{\partial U_0}{\partial \tau} = \frac{\partial^2 U_0}{\partial \theta^2} + U_0(1 - U_0) = \frac{\partial^2 U_0}{\partial \theta^2} + U_0 - U_0^2$$

$$p^2 : \frac{\partial U_2}{\partial \tau} = \frac{\partial^2 U_1}{\partial \theta^2} - U_0 U_1 + U_1(1 - U_0) = \frac{\partial^2 U_1}{\partial \theta^2} + U_1 - 2U_0 U_1$$

$$p^3: \frac{\partial U_3}{\partial \tau} = \frac{\partial^2 U_2}{\partial \theta^2} + U_2 - U_1^2 - 2U_0 U_2$$

From inverse Aboodh transform terms,

$$U_0 = U(\theta, 0) = \frac{1}{(1 + e^{\frac{\theta}{\sqrt{6}}})^2}$$

$$U_1 = A^{-1}[\frac{1}{v} A(\frac{\partial^2 U_0}{\partial \theta^2} + U_0 - U_0^2)] = \frac{5e^{\frac{\theta}{\sqrt{6}}} \tau^1}{3(1 + e^{\frac{\theta}{\sqrt{6}}})^3 1!}$$

$$U_2 = A^{-1}[\frac{1}{v} A(\frac{\partial^2 U_1}{\partial \theta^2} + U_1 - 2U_0 U_1)] = \frac{25e^{\frac{\theta}{\sqrt{6}}} (2e^{\frac{\theta}{\sqrt{6}}} - 1) \tau^2}{18(1 + e^{\frac{\theta}{\sqrt{6}}})^4 (2)!}$$

$$U_3 = A^{-1}[\frac{1}{v} A(\frac{\partial^2 U_2}{\partial \theta^2} + U_2 - U_1^2 - 2U_0 U_2)] = \frac{250e^{\frac{\theta}{\sqrt{6}}} (4(e^{\frac{\theta}{\sqrt{6}}})^2 + 1 - 7e^{\frac{\theta}{\sqrt{6}}}) t^3}{216(1 + e^{\frac{\theta}{\sqrt{6}}})^5 (3)!}$$

Approximate solution is given by,

$$U(\theta, \tau) = \sum_{i=0}^{i=\infty} U_i$$

$$U(x, t) = \frac{1}{(1 + e^{\frac{\theta}{\sqrt{6}}})^2} + \frac{5e^{\frac{\theta}{\sqrt{6}}} \tau^1}{3(1 + e^{\frac{\theta}{\sqrt{6}}})^3 1!} + \frac{25e^{\frac{\theta}{\sqrt{6}}} (2e^{\frac{\theta}{\sqrt{6}}} - 1) \tau^2}{18(1 + e^{\frac{\theta}{\sqrt{6}}})^4 (2)!}$$

$$+ \frac{250e^{\frac{\theta}{\sqrt{6}}} (4(e^{\frac{\theta}{\sqrt{6}}})^2 + 1 - 7e^{\frac{\theta}{\sqrt{6}}}) t^3}{216(1 + e^{\frac{\theta}{\sqrt{6}}})^5 (3)!} + ...$$

Classical NWSE has the solution

$$U(\theta, \tau) = \frac{1}{(1 + e^{\frac{\theta}{\sqrt{6}} - \frac{5\tau}{6}})^2}$$

Which is similar solution to the ADM [20], VIM [12] and HPM [14].

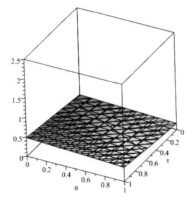

Figure 9.5   Approximate solution for E.g. 3

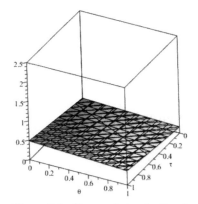

Figure 9.6   Exact solution for E.g. 3

9.5 Conclusion

Here we have successfully applied the new approach Aboodh transform homotopy perturbation method (ATHPM) to find the analytical solution to the NWSE model. Aboodh transformation is exceptionally successful than different strategies examined in the literature. It is limited to overcoming nonlinear issues. Consequently, HPM is presented for the nonlinear terms within the test problems. The proposed graphs and analytical evaluation verify the effectiveness of the advanced approach. We notified a high agreement with the obtained results, and the exact solution was obtained. This hybrid technique appears relevant to linear and nonlinear differential equations. It has been concluded that the arrangement for the NWSE condition empowers us to consider different nonlinear problems. The results were compared with other analytical solutions to the equations. The justification of ATHPM is its straightforwardness and capacity to grant the arrangement to the nonlinear systems with high precision. Future researchers can use this ATHPM method to solve nonlinear DE with high accuracy.

References

[1] Aboodh, Khalid Suliman."The New Integral Transform'Aboodh Transform" GlobalJournal of Pure and Applied Mathematics 9.1 (2013): 35-43.

[2] Aboodh, Khalid Suliman. "Application of new transform "Aboodh Transform" to partialdifferential equations." Global Journal of Pure and Applied Mathematics 10.2 (2014):249-254.

[3] He, Ji-Huan."A coupling method of a homotopy technique and a perturbation techniquefor non-linear problems." International journal of non-linear mechanics 35.1 (2000): 37-43.

[4] He, Ji-Huan. "Homotopy perturbation method: a new nonlinear analytical technique." Applied Mathematics and computation 135.1 (2003): 73-79.

[5] He, Ji-Huan. "Application of homotopy perturbation method to nonlinear wave equations." Chaos, Solitons Fractals 26.3 (2005): 695-700.

[6] Olubanwo, Oludapo Omotola, Olutunde Samuel Odetunde, and Adetoro Temitope Talabi. "Aboodh Homotopy Perturbation Method of solving Burgers Equation." Asian Journal of Applied Sciences 7.2 (2019).

[7] Prakash, Amit, and Manoj Kumar. "He's variational iteration method for the solution of nonlinear Newell–Whitehead–Segel equation." J. Appl. Anal. Comput 6.3 (2016): 738-748.

[8] Pue-On, P. "Laplace Adomian decomposition method for solving Newell–Whitehead–Segel equation." Applied Mathematical Sciences 7.132 (2013): 6593-6600.

[9] Nourazar, S. Salman, Mohsen Soori, and Akbar Nazari-Golshan. "On the exact solution of Newell-Whitehead-Segel equation using the homotopy perturbation method." arXiv preprint arXiv:1502.08016 (2015).

[10] Kumar, Devendra, and Ram Prakash Sharma."Numerical approximation of Newell-Whitehead-Segel equation of fractional order." Nonlinear Engineering 5.2 (2016).

[11] Mahgoub, M. A. "Homotopy Perturbation Method for Solving Newell-Whitehead–Segel Equation, Adv. Theor." Appl. Math 11 (2016): 399-406.

[12] Nadeem, M., S. W. Yao, and N. Parveen. "Solution of Newell-Whitehead-Segel equation by variational iteration method with He's polynomials." Journal of Mathematics and Computer Science 20.1 (2020): 21-29.

[13] Prakash, Amit, Manish Goyal, and Shivangi Gupta. "Fractional variational iteration method for solving time-fractional Newell-Whitehead-Segel equation." Nonlinear engineering 8.1 (2019): 164-171.

[14] Manaa, Saad A. "An approximate solution to the Newell-Whitehead equation by Adomian decomposition method." AL-Rafidain Journal of Computer Sciences and Mathematics 8.1 (2011): 171-180.

[15] Prakash, Amit, and Vijay Verma. "Numerical method for fractional model of Newell-Whitehead-Segel equation." Frontiers in Physics 7 (2019): 15.

[16] Nadeem, Muhammad, Ji-Huan He, and Asad Islam. "The homotopy perturbation method for fractional differential equations: part 1 Mohand transform." International Journal of Numerical Methods for Heat Fluid Flow (2021).

10

Transmission and Control of Droplet Infection from Exotic to Native Population: A Mathematical Model

Chanda Purushwani and Hema Purushwani

Department of Mathematics, School of Science, ITM University, Gwalior 474010, MP, India
E-mail: chandapurushwani1987@outlook.com;
hemapurushwani1985@outlook.com

Abstract

Various infections spread to individuals from an exotic population to native ones. There are limited measures taken by different people to prevent the infection caused by droplets due to limited knowledge. In the present chapter, a mathematical model is designed to study the disease dynamics that spread from exotic populations to native population by droplets. In this model, it is assumed that the native population uses masks as a control measure to prevent disease transmission. The Linear stability analysis and nonlinear stability analysis around disease free and endemic equilibrium points of the model have been determined. The basic reproductive number R_0 has been derived to examine the existence of equilibrium points. It is shown that if the value of a basic reproductive number is less than one then the model is locally stable around disease-free equilibrium point. Similarly, if a value of the basic reproductive number is more than one then the model is locally stable around an endemic free equilibrium point. The numerical simulations of the model show that the droplet infection will become endemic with time or will die out either due to the use of masks or due to the absence of a host.

Keywords: Equilibrium points, Basic Reproduction number, Stability Analysis, Sensitivity analysis, Optimal control problem.

10.1 Introduction

The different diseases that are infectious in nature cause a deleterious impact on human life and are usually caused by carriers that may be human beings themselves as well from other areas which are not found locally have been referred to as exotic in this study. They can arrive by accident or for some purpose in the new environment. There is a huge number of exotic plants, animals and human being species that came from a different native areas are damagingly affecting the native population in an ecosystem [1–8]. Migration is a law of nature; different organisms travel from one place to other and cause their impact on the others. In this chapter, we have considered that infection is spread from exotic population to native population by droplets from exotic to native population by different methods. The infection may spread from exotic to native people, by inhalation of air droplets or aerosols that are contaminated with pathogens from infected persons. The problems are further aggravated by the mass movement of large numbers of people who are infected as currently there is no proper surveillance system to detect infections. The presence of such individuals in a different environment will play a vital role in the spread of contagious disease as the atmosphere of persons is an indirect means of disease spread [9–12].

To detect the prediction of emerging communicable diseases from exotic to native populations various methods are developed, one of them is mathematical modeling. The pathogens like viruses and bacteria may give rise to West Nile virus, Influenza, Measles, Salmonella, Anthrax, Cholera, and Chlamydia if they are present in an environment where native people are inhibited. Disease may be propagated directly from one human being to another human being by body secretion (e.g. Chlamydia) and respiratory droplets (e.g. Measles). Most of the infections can be controlled by antibiotics, antiviral drugs, vaccines, and increased sanitation or behavioral changes like using a mask while coughing or sneezing [6, 11, 12].

In recent years, the different diseases are transmitting at a very rapid rate as land transport, sea transport and air transport channels continue to expand in reach and as such pathogens and vectors can transmit very fast will lead to vector-borne pathogens importation, infection disease pandemics and vector invasion events [14]. One of the highly contagious diseases that spread from one country to other parts of the world is Swine flu which was first reported in Mexico and subsequently was reported from other parts of the world. Swine flu transmits from human beings to human beings during contact, touch, and droplets (carrying virus and bacteria) from person to environmental surfaces by coughing and sneezing [10, 13]. The other infectious disease is Plague is that spread by Yersinia pestis bacteria, estimated 75–200 million people in Europe were infected by these epidemics.

Several types of researchers like Misra, O.P. et al. 2009, Braian, D. et al. 2015, and Li, M.T. et al. 2017 have suggested mathematical models understand the population dynamics of infection agents and their impact on the exposed population [1, 8, 10]. However, in this chapter, we argue that exotic populations may infect native people by droplet infection if they come in contact with them as most of the time, people who travel to other parts of the world communicate/contact with almost all strata of the population. Thus, it is believed that the exotic population who travel to different regions has to contact with the native populations and may become victims of infecting agents to whom they are not exposed in their native habitats.

This chapter is prepared in the following manner; Section 10.2 includes Basic assumptions and formulation of the model to analyze the effect of droplet infection on the native population. In section 10.3, we have found Disease free equilibrium point E_0 and Basic reproductive R_o. Stability analysis (Linear and Non linear) around disease free equilibrium points has been done in section 10.4. Condition for the existence of endemic points has been derived in section 10.5. Section 10.6 contains stability (Linear and Non linear) around the endemic equilibrium point. Sensitivity analysis of basic reproduction number and state variables has been performed in section 10.7. In section 10.8, an optimal control problem is solved to minimize in infected population, pathogens, and related costs. Numerical simulation for the model is done as well as relevant sketches have been plotted in section 10.9. In section 10.10, a Conclusion with important suggestions has been provided on the basis of sensitivity indices for the study.

10.2 Basic Assumptions and Formulation of the Model

The human population (N) taken in the study is consist of the native population (N_1) and the exotic population (N_2). We assumed N_1 is distributed into three categories: Susceptible (S_1), Infected (I_1) and Recovered (R_1) from native population. Similarly, N_2 is distributed into three categories: Susceptible (S_2), Infected (I_2) and Recovered (R_2) from exotic population. P is represented as a pathogen class created by infectious exotic human being's droplets. Δ_1 and Δ_2 are the susceptible human being's (S_1 and S_2) recruitment rate, respectively. The susceptible population S_1 and S_2 reduce due to infection by pathogens P at the rate $(1-a)\beta_1 S_1 P$ and

$(1 - a)\beta_2 S_2 P$, respectively. Infected population I_1 and I_2 increase with the rate of $(1 - a)\beta_1 S_1 P$ and $(1 - a)\beta_2 S_2 P$, respectively and infected individuals I_1 and I_2 recover with the rate $\alpha_1 I_1$ and $\alpha_2 I_2$, respectively there is decrease in I_1 and I_2 with the death rate σ_1 and σ_2, respectively due to infection. Recovered class R_1 and R_2 increase by $\alpha_1 I_1$ and $\alpha_2 I_2$, respectively and Class R_1 and R_2 decrease at the rate of $\vartheta_1 I_1$ and $\vartheta_2 I_2$, respectively due to immunity so these populations becomes susceptible again. All human populations have a natural death rate μ. Since infection spread through pathogens produced by droplets of an infected persons. In the model, we have supposed that infected human beings I_2 produce pathogens population P at the rate η and due to natural death rate δ, these pathogens moderate.

In view of above to explain transmission mechanism, following mathematical model is designed [13–17];

$$\frac{dS_1}{dt} = \Delta_1 - (1 - a)\beta_1 S_1 P - \mu S_1 + \vartheta_1 R_1 \tag{10.1}$$

$$\frac{dI_1}{dt} = (1 - a)\beta_1 S_1 P - (\alpha_1 + \mu + \sigma_1) I_1 \tag{10.2}$$

$$\frac{dR_1}{dt} = \alpha_1 I_1 - (\mu + \vartheta_1) R_1 \tag{10.3}$$

$$\frac{dS_2}{dt} = \Delta_2 - (1 - a)\beta_2 S_2 P - \mu S_2 + \vartheta_2 R_2 \tag{10.4}$$

$$\frac{dI_2}{dt} = (1 - a)\beta_2 S_2 P - (\alpha_2 + \mu + \sigma_2) I_2 \tag{10.5}$$

$$\frac{dR_2}{dt} = \alpha_2 I_2 - (\mu + \vartheta_2) R_2 \tag{10.6}$$

$$\frac{dP}{dt} = \eta I_2 - \delta P \tag{10.7}$$

with initial conditions

$$S_i(0) = S_{i0} > 0, \quad I_i(0) = I_{i0} > 0, \quad R_i(0) = R_{i0} > 0 \text{ and } P(0) = P_0 > 0 \text{ for}$$

$$(i = 1, 2) \tag{10.8}$$

Description of parameters and variable is given below;
Δ_1 = Native human being's recruitment rate.
Δ_2 = Exotic human being's recruitment rate.
β_i = Rate of infection transmits to susceptible human being's through pathogens $(i = 1, 2)$.
μ = Human being's natural death rate.
ϑ_i = Temporary recovery rate from recovered population to susceptible population

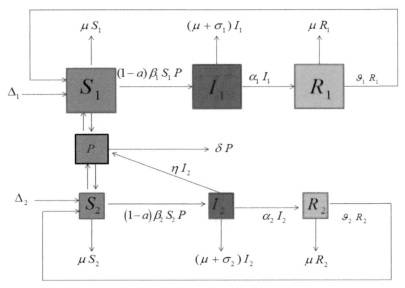

Figure 10.1   Disease propagation sketch of designed model

$(i = 1, 2)$.
α_i = Permanent recovery rate from infected population to recovered population $(i = 1, 2)$.
σ_i = Human being's disease caused death rate $(i = 1, 2)$.
η = Pathogen's produce rate by infected human being.
δ = Pathogen's natural death rate.
a = Efficiency of mask used by population.

Disease propagation sketch of designed model is highlighted in Figure 10.1 to show the infection transmission.

Suppose $N_1 = S_1 + I_1 + R_1$ and $N_2 = S_2 + I_2 + R_2$ are total native and exotic population, respectively.

Now from the model system (10.1) to (10.7), we have;

$$\frac{dN_1}{dt} = \Delta_1 - \mu N_1 - \sigma_1 I_1$$

$$\frac{dN_1}{dt} \leq \Delta_1 - \mu N_1$$

$$\frac{dN_1}{dt} + \mu N_1 \leq \Delta_1$$

$$\lim_{t \to \infty} N_1 \leq \frac{\Delta_1}{\mu} \tag{10.9}$$

and again from the model system (10.1) to (10.7), we have;

$$\frac{dN_2}{dt} = \Delta_2 - \mu N_2 - \sigma_2 I_2$$

$$\frac{dN_2}{dt} \leq \Delta_2 - \mu N_2$$

$$\frac{dN_2}{dt} + \mu N_2 \leq \Delta_2$$

$$\lim_{t \to \infty} N_2 \leq \frac{\Delta_2}{\mu} \tag{10.10}$$

Further, from the model system (10.1) to (10.7), we have;

$$\frac{dP}{dt} = \eta I_2 - \delta P$$

$$\frac{dP}{dt} \leq \frac{\eta \Delta_2}{\mu} - \delta P$$

$$\frac{dP}{dt} + \delta P \leq \frac{\eta \Delta_2}{\mu}$$

$$\lim_{t \to \infty} P \leq \frac{\eta \Delta_2}{\mu \delta} \tag{10.11}$$

Consequently all the results of model system (10.1) to (10.7) will exist in the locale

$$\tau = \left\{ \left(\bar{S}_1, \bar{I}_1, \bar{R}_1, \bar{S}_2, \bar{I}_2, \bar{R}_2, \bar{P} \right) : \bar{S}_1, \bar{I}_1, \bar{R}_1, \bar{S}_2, \bar{I}_2, \bar{R}_2, \bar{P} \geq 0 : \right.$$

$$\left. \left(\bar{S}_1 + \bar{I}_1 + \bar{R}_1 \right) \leq \frac{\Delta_1}{\mu}, \left(\bar{S}_2 + \bar{I}_2 + \bar{R}_2 \right) \leq \frac{\Delta_2}{\mu} \text{ and } \bar{P} \leq \frac{\eta \Delta_2}{\mu \delta} \right\} \tag{10.12}$$

Hence evidently τ is a positive dense invariant locale in R_7^+

10.3 Disease Free Equilibrium Point and Basic Reproduction Number

Disease free State indicates there is no disease in the context. Thus model system (10.1) to (10.7) has disease free equilibrium point $E_0 = \left(\bar{S}_1, \bar{I}_1, \bar{R}_1, \bar{S}_2, \bar{I}_2, \bar{R}_2, \bar{P} \right) = \left(\frac{\Delta_1}{\mu}, 0, 0, \frac{\Delta_2}{\mu}, 0, 0, 0 \right)$. in the region τ.

Now we will evaluate basic reproductive number R_0 of model system (10.1) to (10.7) using Next generation matrix method [18–23] as follows;

$$f_1 (I_1, I_2, P) = (1 - a) \beta_1 S_1 P, \ f_2 (I_1, I_2, P) = (1 - a) \beta_2 S_2 P \text{ and}$$
$$f_3 (I_1, I_2, P) = 0$$

$$V_1(I_1, I_2, P) = (\alpha_1 + \mu + \sigma_1)I_1, \ V_2(I_1, I_2, P) = (\alpha_2 + \mu + \sigma_2)I_2 \text{ and}$$
$$V_3(I_1, I_2, P) = \delta P - \eta I_2$$

Therefore

$$f = \begin{bmatrix} \frac{\partial f_1}{\partial I_1} & \frac{\partial f_1}{\partial I_2} & \frac{\partial f_1}{\partial P} \\ \frac{\partial f_2}{\partial I_1} & \frac{\partial f_2}{\partial I_2} & \frac{\partial f_2}{\partial P} \\ \frac{\partial f_3}{\partial I_1} & \frac{\partial f_3}{\partial I_2} & \frac{\partial f_3}{\partial P} \end{bmatrix} = \begin{bmatrix} 0 & 0 & (1-a)\beta_1 S_1 \\ 0 & 0 & (1-a)\beta_2 S_2 \\ 0 & 0 & 0 \end{bmatrix}$$

and

$$V = \begin{bmatrix} \frac{\partial V_1}{\partial I_1} & \frac{\partial V_1}{\partial I_2} & \frac{\partial V_1}{\partial P} \\ \frac{\partial V_2}{\partial I_1} & \frac{\partial V_2}{\partial I_2} & \frac{\partial V_2}{\partial P} \\ \frac{\partial V_3}{\partial I_1} & \frac{\partial V_3}{\partial I_2} & \frac{\partial V_3}{\partial P} \end{bmatrix} = \begin{bmatrix} (\alpha_1 + \mu + \sigma_1) & 0 & 0 \\ 0 & (\alpha_2 + \mu + \sigma_2) & 0 \\ 0 & -\eta & \delta \end{bmatrix}$$

$$V^{-1} = \begin{bmatrix} \frac{1}{(\alpha_1 + \mu + \sigma_1)} & 0 & 0 \\ 0 & \frac{1}{(\alpha_2 + \mu + \sigma_2)} & 0 \\ 0 & \frac{\eta}{\delta(\alpha_2 + \mu + \sigma_2)} & \frac{1}{\delta} \end{bmatrix}$$

$$fV^{-1} = \begin{bmatrix} 0 & \frac{(1-a)\beta_1 \bar{S}_1 \eta}{\delta(\alpha_2 + \mu + \sigma_2)} & \frac{(1-a)\beta_1 \bar{S}_1}{\delta} \\ 0 & \frac{(1-a)\beta_2 \bar{S}_2 \eta}{\delta(\alpha_2 + \mu + \sigma_2)} & \frac{(1-a)\beta_2 \bar{S}_2}{\delta} \\ 0 & 0 & 0 \end{bmatrix}$$

Now the characteristic equation of above matrix is given below:

$$\lambda^2 \left(\lambda - \frac{(1-a)\beta_2 \bar{S}_2 \eta}{\delta(\alpha_2 + \mu + \sigma_2)} \right) = 0$$

The dominating Eigen value of above matrix is $\frac{(1-a)\beta_2 \bar{S}_2 \eta}{\delta(\alpha_2 + \mu + \sigma_2)}$ which is known as basic reproduction number Thus, $R_0 = \frac{(1-a)\beta_2 \Delta_2 \eta}{\delta\mu(\alpha_2 + \mu + \sigma_2)}$

10.4 Stability Around the Disease Free Equilibrium Point

After implementing the following transforms on model system (10.1) to (10.7), we get;

$$S_1 = \frac{\Delta_1}{\mu} + x_1, \ I_1 = x_2, \ R_1 = x_3, \ S_2 = \frac{\Delta_2}{\mu} + x_4, \ I_2 = x_5, \ R_2 = x_6 \text{ and } P = x_7$$

$$\frac{dx_1}{dt} = -(1-a)\beta_1 \left(\frac{\Delta_1}{\mu} + x_1 \right) x_7 - \mu x_1 + \vartheta_1 x_3$$

$$\frac{dx_2}{dt} = (1-a)\beta_1 \left(\frac{\Delta_1}{\mu} + x_1 \right) x_7 - (\alpha_1 + \mu + \sigma_1) x_2$$

$$\frac{dx_3}{dt} = \alpha_1 x_2 - (\mu + \vartheta_1) x_3$$

$$\frac{dx_4}{dt} = -(1-a)\beta_2 \left(\frac{\Delta_2}{\mu} + x_4\right) x_7 - \mu x_4 + \vartheta_2 x_6$$

$$\frac{dx_5}{dt} = (1-a)\beta_2 \left(\frac{\Delta_2}{\mu} + x_4\right) x_7 - (\alpha_2 + \mu + \sigma_2) x_5$$

$$\frac{dx_6}{dt} = \alpha_2 x_5 - (\mu + \vartheta_2) x_6$$

$$\frac{dx_7}{dt} = \eta x_5 - \delta x_7 \tag{10.13}$$

10.4.1 Linear Stability Around the Disease Free Equilibrium Point E_0

To examine linear stability around the disease free equilibrium point E_0 we ignore non linear terms and calculate variational matrix J_0 as follows,

$$\frac{dx_1}{dt} = -\frac{(1-a)\beta_1\Delta_1}{\mu} x_7 - \mu x_1 + \vartheta_1 x_3$$

$$\frac{dx_2}{dt} = \frac{(1-a)\beta_1\Delta_1}{\mu} x_7 - (\alpha_1 + \mu + \sigma_1) x_2$$

$$\frac{dx_3}{dt} = \alpha_1 x_2 - (\mu + \vartheta_1) x_3$$

$$\frac{dx_4}{dt} = -\frac{(1-a)\beta_2\Delta_2}{\mu} x_7 - \mu x_4 + \vartheta_2 x_6$$

$$\frac{dx_5}{dt} = \frac{(1-a)\beta_2\Delta_2}{\mu} x_7 - (\alpha_2 + \mu + \sigma_2) x_5$$

$$\frac{dx_6}{dt} = \alpha_2 x_5 - (\mu + \vartheta_2) x_6$$

$$\frac{dx_7}{dt} = \eta x_5 - \delta x_7$$

The variational matrix of model system (10.1) to (10.7) around E_0 is given as follows;

$$J_0 = \begin{bmatrix} -\mu & 0 & \vartheta_1 & 0 & 0 & 0 & -\frac{(1-a)\beta_1\Delta_1}{\mu} \\ 0 & -(\alpha_1+\mu+\sigma_1) & 0 & 0 & 0 & 0 & \frac{(1-a)\beta_1\Delta_1}{\mu} \\ 0 & \alpha_1 & -(\mu+\vartheta_1) & 0 & 0 & 0 & 0 \\ 0 & 0 & 0 & -\mu & 0 & \vartheta_2 & -\frac{(1-a)\beta_2\Delta_2}{\mu} \\ 0 & 0 & 0 & 0 & -(\alpha_2+\mu+\sigma_2) & 0 & \frac{(1-a)\beta_2\Delta_2}{\mu} \\ 0 & 0 & 0 & 0 & \alpha_2 & -(\mu+\vartheta_2) & 0 \\ 0 & 0 & 0 & 0 & \eta & 0 & -\delta \end{bmatrix}$$

The characteristic equation of J_0 is given below;

$$(\mu + \lambda)^2 (\mu + \vartheta_1 + \lambda)(\mu + \vartheta_2 + \lambda)(\alpha_1 + \mu + \sigma_1 + \lambda)$$

$$\left(\lambda^2 + (\alpha_2 + \mu + \sigma_2 + \delta)\lambda + (\alpha_2 + \mu + \sigma_2)\delta - (1 - a)\frac{\beta_2 \eta \Delta_2}{\mu}\right) = 0 \quad (10.14)$$

This implies,

$$(\mu + \lambda)^2 (\mu + \vartheta_1 + \lambda)(\mu + \vartheta_2 + \lambda)(\alpha_1 + \mu + \sigma_1 + \lambda)$$

$$\lambda^2 + (\alpha_2 + \mu + \sigma_2 + \delta)\lambda + (\alpha_2 + \mu + \sigma_2)\delta (1 - R_0) = 0$$

It can be easily observed from Descartes' rule of sign and variational matrix method that disease free equilibrium point E_0 is linearly asymptotically stable if $R_0 < 1$ however, the disease free equilibrium point E_0 is unstable if $R_0 > 1$.

10.4.2 Non Linear Stability Around the Disease Free Equilibrium Point E_0

Assume positive definite function U_1 which is given below;

$$U_1 = \frac{1}{2}\left(H_1 x_1^2 + H_2 x_2^2 + H_3 x_3^2 + H_4 x_4^2 + H_5 x_5^2 + H_6 x_6^2 + H_7 x_7^2\right)$$
$$(10.15)$$

Now using system (13) in $\frac{dU_1}{dt}$ we get;

$$
\begin{aligned}
\frac{dU_1}{dt} = & H_1\left(-(1-a)\beta_1\left(\frac{\Delta_1}{\mu} + x_1\right)x_1 x_7 - \mu x_1^2 + \vartheta_1 x_1 x_3\right) \\
& + H_2\left((1-a)\beta_1\left(\frac{\Delta_1}{\mu} + x_1\right)x_2 x_7 - (\alpha_1 + \mu + \sigma_1)x_2^2\right) \\
& + H_3\left(\alpha_1 x_2 x_3 - (\mu + \vartheta_1)x_3^2\right) \\
& + H_4\left(-(1-a)\beta_2\left(\frac{\Delta_2}{\mu} + x_4\right)x_4 x_7 - \mu x_4^2 + \vartheta_2 x_4 x_6\right) \\
& + H_5\left((1-a)\beta_2\left(\frac{\Delta_2}{\mu} + x_4\right)x_5 x_7 - (\alpha_2 + \mu + \sigma_2)x_5^2\right) \\
& + H_6\left(\alpha_2 x_5 x_6 - (\mu + \vartheta_2)x_6^2\right) + H_7\left(\eta x_5 x_7 - \delta x_7^2\right)
\end{aligned}
$$

Now using the inequality, $\pm 2xy \le x^2 + y^2$ on the R.H.S. of $\frac{dU_1}{dt}$, we get;

$$\frac{dU_1}{dt} \le H_1\left(\frac{(1-a)\beta_1 \Delta_1}{2\mu}\left(x_1^2 + x_7^2\right) - \mu x_1^2 + \vartheta_1 x_1 x_3\right)$$

$$+ H_2 \left(\frac{(1-a)\,\beta_1\Delta_1}{2\mu} \left(x_2{}^2 + x_7{}^2 \right) - (\alpha_1 + \mu + \sigma_1)\,x_2{}^2 \right)$$

$$+ H_3 \left(\alpha_1 x_2 x_3 - (\mu + \vartheta_1)\,x_3{}^2 \right)$$

$$+ H_4 \left(\frac{(1-a)\,\beta_2\Delta_2}{2\mu} \left(x_4{}^2 + x_7{}^2 \right) - \mu x_4{}^2 + \vartheta_2 x_4 x_6 \right)$$

$$+ H_5 \left(\frac{(1-a)\,\beta_2\Delta_2}{2\mu} \left(x_5{}^2 + x_7{}^2 \right) - (\alpha_2 + \mu + \sigma_2)\,x_5{}^2 \right)$$

$$+ H_6 \left(\alpha_2 x_5 x_6 - (\mu + \vartheta_2)\,x_6{}^2 \right) + H_7 \left(\eta x_5 x_7 - \delta x_7{}^2 \right)$$

On arranging above inequality we get;

$$\frac{dU_1}{dt} \leq - \left[\left(\mu - \frac{(1-a)\,\beta_1\Delta_1}{2\mu} \right) H_1 x_1{}^2 \right.$$

$$- \vartheta_1 H_1 x_1 x_3 + \left((\alpha_1 + \mu + \sigma_1) - \frac{(1-a)\,\beta_1\Delta_1}{2\mu} \right) H_2 x_2{}^2$$

$$- \alpha_1 H_3 x_2 x_3 + (\mu + \vartheta_1) H_3 x_3{}^2 + \left(\mu - \frac{(1-a)\,\beta_2\Delta_2}{2\mu} \right) H_4 x_4{}^2$$

$$- \vartheta_2 H_4 x_4 x_6 + \left((\alpha_2 + \mu + \sigma_2) - \frac{(1-a)\,\beta_2\Delta_2}{2\mu} \right) H_5 x_5{}^2$$

$$(\mu + \vartheta_2) H_6 x_6{}^2 - \alpha_2 H_6 x_5 x_6 - \eta H_7 x_5 x_7 + (\delta H_7$$

$$\left. - \left(\frac{(1-a)\,\beta_1\Delta_1}{2\mu} (H_1 + H_2) + \frac{(1-a)\,\beta_2\Delta_2}{2\mu} (H_4 + H_5) \right) \right) x_7{}^2 \right]$$

This implies;

$$\frac{dU_1}{dt} \leq - \left[b_{11} x_1{}^2 - b_{13} x_1 x_3 + b_{22} x_2{}^2 - b_{23} x_2 x_3 + b_{33} x_3{}^2 + b_{44} x_4{}^2 \right.$$

$$\left. - b_{46} x_4 x_6 + b_{55} x_5{}^2 + b_{66} x_6{}^2 - b_{56} x_5 x_6 - b_{57} x_5 x_7 + b_{77} x_7{}^2 \right]$$

Where

$$b_{11} = \left(\mu - \frac{(1-a)\,\beta_1\Delta_1}{2\mu} \right) H_1,\, b_{13}\vartheta_1 H_1,$$

$$b_{22} = \left((\alpha_1 + \mu + \sigma_1) - \frac{(1-a)\,\beta_1\Delta_1}{2\mu} \right) H_2,$$

$$b_{23} = \alpha_1 H_3,\, b_{33} = (\mu + \vartheta_1) H_3,\, b_{44} = \left(\mu - \frac{(1-a)\,\beta_2\Delta_2}{2\mu} \right) H_4,\, b_{46} = \vartheta_2 H_4,$$

$$b_{55} = \left((\alpha_2 + \mu + \sigma_2) - \frac{(1-a)\,\beta_2\Delta_2}{2\mu} \right) H_5,\, b_{66} = (\mu + \vartheta_2) H_6,\, b_{56} = \alpha_2 H_6,$$

$$b_{57} = \eta H_7, b_{77} = \delta H_7 - \left(\frac{(1-a)\beta_1 \Delta_1}{2\mu} (H_1 + H_2) \right.$$

$$+ \left. \frac{(1-a)\beta_2 \Delta_2}{2\mu} (H_4 + H_5) \right)$$

Again rearranging the inequality, we get;

$$\frac{dU_1}{dt} \le - \left[\left(b_{11}x_1^2 - b_{13}x_1x_3 + \frac{b_{33}}{2}x_3^2 \right) + \left(b_{22}x_2^2 - b_{23}x_2x_3 + \frac{b_{33}}{2}x_3^2 \right) \right.$$

$$+ \left(b_{44}x_4^2 - b_{46}x_4x_6 + \frac{b_{66}}{2}x_6^2 \right)$$

$$+ \left. \left(\frac{b_{55}}{2}x_5^2 - b_{56}x_5x_6 + \frac{b_{66}}{2}x_6^2 \right) + \left(\frac{b_{55}}{2}x_5^2 - b_{57}x_5x_7 + b_{77}x_7^2 \right) \right]$$

$$(10.16)$$

Implementing the Sylvester criteria on the R.H.S. of inequality (10.16); under below mentioned conditions $\frac{dU_1}{dt}$ is negative definite;

$$2 \left(\mu - \frac{(1-a)\beta_1 \Delta_1}{2\mu} \right) (\mu + \vartheta_1) H_3 > \vartheta_1^2 H_1,$$

$$2 \left((\alpha_1 + \mu + \sigma_1) - \frac{(1-a)\beta_1 \Delta_1}{2\mu} \right) (\mu + \vartheta_1) H_2 > \alpha_1^2 H_3,$$

$$2 \left(\mu - \frac{(1-a)\beta_2 \Delta_2}{2\mu} \right) (\mu + \vartheta_2) H_6 > \vartheta_2^2$$

$$H_4, \left((\alpha_2 + \mu + \sigma_2) - \frac{(1-a)\beta_2 \Delta_2}{2\mu} \right) (\mu + \vartheta_2) H_5 > \alpha_2^2 H_6,$$

$$\left\{ 2 \left(\delta H_7 - \left(\frac{(1-a)\beta_1 \Delta_1}{2\mu} (H_1 + H_2) + \frac{(1-a)\beta_2 \Delta_2}{2\mu} (H_4 + H_5) \right) \right) \right.$$

$$\left. \left((\alpha_2 + \mu + \sigma_2) - \frac{(1-a)\beta_2 \Delta_2}{2\mu} \right) H_5 \right\} > (\eta H_7)^2. \qquad (10.17)$$

Now taking $H_1 = H_2 = H_4 = H_5 = 1$ and choosing H_3, H_6, H_7.

$$\frac{2\mu \vartheta_1^2}{(2\mu^2 - (1-a)\beta_1 \Delta_1)(\mu + \vartheta_1)} < H_3 < \frac{(2\mu(\alpha_1 + \mu + \sigma_1) - (1-a)\beta_1 \Delta_1)(\mu + \vartheta_1)}{\alpha_1^2 \mu},$$

$$\frac{\mu \vartheta_2^2}{(2\mu^2 - (1-a)\beta_2 \Delta_2)(\mu + \vartheta_2)} < H_6 < \frac{(2\mu(\alpha_2 + \mu + \sigma_2) - (1-a)\beta_2 \Delta_2)(\mu + \vartheta_2)}{2\mu \alpha_2^2},$$

$$H_7 > \frac{(1-a)(\beta_1 \Delta_1 + \beta_2 \Delta_2)}{\mu \delta}.$$

Implementing Sylvester's criteria, under below mentioned conditions $\frac{dU_1}{dt}$ is negative definite;

$$2\mu^2 > (1-a)\beta_1 \Delta_1,$$

$$2\mu\left(\alpha_1 + \mu + \sigma_1\right) > \left(1 - a\right)\beta_1\Delta_1,$$

$$2\mu^2 > \left(1 - a\right)\beta_2\Delta_2$$

and $2\mu\left(\alpha_2 + \mu + \sigma_2\right) > \left(1 - a\right)\beta_2\Delta_2$.

Consequently, the disease free equilibrium point E_0 is non linearly stable under above mentioned conditions otherwise unstable (using Lyapunov's theorem).

10.5 Existence of Endemic Equilibrium Point

The endemic equilibrium point $E_1 = \left(\hat{S}_1, \hat{I}_1, \hat{R}_1, \hat{S}_2, \hat{I}_2, \hat{R}_2, \hat{P}\right)$ in the region τ is obtained by Putting

$$\frac{d\hat{S}_1}{dt} = \frac{d\hat{I}_1}{dt} = \frac{d\hat{R}_1}{dt} = \frac{d\hat{S}_2}{dt} = \frac{d\hat{I}_2}{dt} = \frac{d\hat{R}_2}{dt} = \frac{d\hat{P}}{dt} = 0 \ ,$$

this gives

$$\hat{S}_1 = \frac{\delta(\alpha_1 + \mu + \sigma_1)}{(1-a)\beta_1\eta}\left(\frac{\hat{I}_1}{\hat{I}_2}\right),$$

$$\hat{S}_2 = \frac{(\alpha_2 + \mu + \sigma_2)\,\delta}{(1-a)\,\eta\beta_2},$$

$$\hat{R}_1 = \left(\frac{\alpha_1}{\mu + \vartheta_1}\right)\hat{I}_1,$$

$$\hat{R}_2 = \left(\frac{\alpha_2}{\mu + \vartheta_2}\right)\hat{I}_2,$$

$$\hat{P} = \frac{\eta\hat{I}_2}{\delta},$$

$$\hat{I}_1 = \frac{\Delta_1}{\left(\frac{\mu\delta(\alpha_1+\mu+\sigma_1)}{(1-a)\beta_1\eta\hat{I}_2} + \frac{\mu\alpha_1}{(\mu+\vartheta_1)} + (\mu + \sigma_1)\right)},$$

$$\hat{I}_2 = \frac{\delta\mu(\alpha_2 + \mu + \sigma_2)\,(\mu + \vartheta_2)\left(\frac{(1-a)\beta_2\Delta_2\eta}{\delta\mu(\alpha_2+\mu+\sigma_2)} - 1\right)}{\left((\mu + \sigma_2)\,(\mu + \vartheta_2) + \mu\alpha_2\right)(1 - a)\,\eta\beta_2}$$

$$= \frac{\delta\mu(\alpha_2 + \mu + \sigma_2)\,(\mu + \vartheta_2)\,(R_0 - 1)}{\left((\mu + \sigma_2)\,(\mu + \vartheta_2) + \mu\alpha_2\right)(1 - a)\,\eta\beta_2}.$$

Consequently, when $R_0 > 1$ $\hat{I}_2$ is positive this implies all $\hat{S}_1, \hat{I}_1, \hat{R}_1, \hat{S}_2, \hat{R}_2, \hat{P}$ will be positive so $E_1 = \left(\hat{S}_1, \hat{I}_1, \hat{R}_1, \hat{S}_2, \hat{I}_2, \hat{R}_2, \hat{P}\right)$ exist when $R_0 > 1$ in region τ

10.6 Stability Around the Endemic Equilibrium Point

Implementing the following transforms on model system (10.1) to (10.7), we get;

$$S_1 = \hat{S}_1 + y_1, \ I_1 = \hat{I}_1 + y_2, \ R_1 = \hat{R}_1 + y_3, \ S_2 = \hat{S}_2 + y_4, \ I_2 = \hat{I}_2 + y_5,$$
$$R_2 = \hat{R}_2 + y_6 \ \text{and} \ P = \hat{P} + y_7.$$

$$\frac{dy_1}{dt} = -\left(1 - a\right)\beta_1\left(\hat{S}_1 y_7 + \hat{P} y_1 + y_1 y_7\right) - \mu y_1 + \vartheta_1 y_3$$

$$\frac{dy_2}{dt} = \left(1 - a\right)\beta_1\left(\hat{S}_1 y_7 + \hat{P} y_1 + y_1 y_7\right) - \left(\alpha_1 + \mu + \sigma_1\right) y_2$$

$$\frac{dy_3}{dt} = \alpha_1 y_2 - \left(\mu + \vartheta_1\right) y_3$$

$$\frac{dy_4}{dt} = -\left(1 - a\right)\beta_2\left(\hat{S}_2 y_7 + \hat{P} y_4 + y_4 y_7\right) - \mu y_4 + \vartheta_2 y_6$$

$$\frac{dy_5}{dt} = \left(1 - a\right)\beta_2\left(\hat{S}_2 y_7 + \hat{P} y_4 + y_4 y_7\right) - \left(\alpha_2 + \mu + \sigma_2\right) y_5$$

$$\frac{dy_6}{dt} = \alpha_2 y_5 - \left(\mu + \vartheta_2\right) y_6$$

$$\frac{dy_7}{dt} = \eta y_5 - \delta y_7 \tag{10.18}$$

10.6.1 Linear Stability Around the Endemic Equilibrium Point E_1

To observe linear stability around the endemic equilibrium, point E_1 we ignore non linear terms and calculate variational matrix J_1 as follows,

$$\frac{dy_1}{dt} = -\left(1 - a\right)\beta_1\left(\hat{S}_1 y_7 + \hat{P} y_1\right) - \mu y_1 + \vartheta_1 y_3$$

$$\frac{dy_2}{dt} = \left(1 - a\right)\beta_1\left(\hat{S}_1 y_7 + \hat{P} y_1\right) - \left(\alpha_1 + \mu + \sigma_1\right) y_2$$

$$\frac{dy_3}{dt} = \alpha_1 y_2 - \left(\mu + \vartheta_1\right) y_3$$

$$\frac{dy_4}{dt} = -\left(1 - a\right)\beta_2\left(\hat{S}_2 y_7 + \hat{P} y_4\right) - \mu y_4 + \vartheta_2 y_6$$

$$\frac{dy_5}{dt} = \left(1 - a\right)\beta_2\left(\hat{S}_2 y_7 + \hat{P} y_4\right) - \left(\alpha_2 + \mu + \sigma_2\right) y_5$$

$$\frac{dy_6}{dt} = \alpha_2 y_5 - \left(\mu + \vartheta_2\right) y_6$$

$$\frac{dy_7}{dt} = \eta y_5 - \delta y_7$$

The variational matrix of model system (10.1) to (10.7) around E_1 is given as follows:

$$J_1 =$$

$$
\begin{bmatrix}
-(1-a)\beta_1\hat{P}-\mu & 0 & \vartheta_1 & 0 & 0 & 0 & -(1-a)\beta_1\hat{S}_1 \\
(1-a)\beta_1\hat{P} & -(\alpha_1+\mu+\sigma_1) & 0 & 0 & 0 & 0 & (1-a)\beta_1\hat{S}_1 \\
0 & \alpha_1 & -(\mu+\vartheta_1) & 0 & 0 & 0 & 0 \\
0 & 0 & 0 & -(1-a)\beta_2\hat{P}-\mu & 0 & \vartheta_2 & -(1-a)\beta_2\hat{S}_2 \\
0 & 0 & 0 & (1-a)\beta_2\hat{P} & -(\alpha_2+\mu+\sigma_2) & 0 & (1-a)\beta_2\hat{S}_2 \\
0 & 0 & 0 & 0 & \alpha_2 & -(\mu+\vartheta_2) & 0 \\
0 & 0 & 0 & 0 & \eta & 0 & -\delta
\end{bmatrix}
$$

The characteristic equation of J_1 is given below;

$$\left(\lambda^3 + a_1\lambda^2 + a_2\lambda + a_3\right)\left(\lambda^4 + b_1\lambda^3 + b_2\lambda^2 + b_3\lambda + b_4\right) = 0 \tag{10.19}$$

where,

$$a_1 = (1-a)\beta_1\hat{P} + \mu + (\alpha_1+\mu+\sigma_1) + (\mu+\vartheta_1),$$

$$a_2 = (\alpha_1+\mu+\sigma_1)(\mu+\vartheta_1) + \left((1-a)\beta_1\hat{P}+\mu\right)((\alpha_1+\mu+\sigma_1)+(\mu+\vartheta_1)),$$

$$a_3 = (1-a)\beta_1\alpha_1\mu\hat{P} + (1-a)\beta_1(\mu+\sigma_1)(\mu+\vartheta_1)\hat{P}$$
$$+\mu(\alpha_1+\mu+\sigma_1)(\mu+\vartheta_1),$$

$$b_1 = \left((1-a)\beta_2\hat{P}+\mu\right) + (\alpha_2+\mu+\sigma_2) + (\mu+\vartheta_2) + \delta,$$

$$b_2 = \left((1-a)\beta_2\hat{P}+\mu\right)((\alpha_2+\mu+\sigma_2)+(\mu+\vartheta_2)+\delta)$$
$$+(\alpha_2+\mu+\sigma_2+\delta)(\mu+\vartheta_2),$$

$$b_3 = (1-a)\beta_2\hat{P}((\alpha_2+\mu+\sigma_2)\delta + \alpha_2\mu + (\mu+\sigma_2)(\mu+\vartheta_2))$$
$$+\left((1-a)\beta_2\hat{P}+\mu\right)(\mu+\vartheta_2)\delta + \mu(\alpha_2+\mu+\sigma_2)(\mu+\vartheta_2),$$

$$b_4 = (1-a)\beta_2\hat{P}\delta(\alpha_2\mu + (\mu+\sigma_2)(\mu+\vartheta_2)).$$

It can be easily observed from Descartes' rule of sign and variational matrix method that endemic equilibrium point E_1 is linearly asymptotically stable if $\hat{P} > 0$ i.e. $R_0 > 1$. $R_0 > 1$ however, the endemic equilibrium point E_1 is unstable if $R_0 < 1$.

10.6.2 Non Linear Stability Around the Endemic Equilibrium Point E_1

Assume a positive definite function U_2 given below;

$$U_2 = \frac{1}{2}\left(G_1y_1{}^2 + G_2y_2{}^2 + G_3y_3{}^2 + G_4y_4{}^2 + G_5y_5{}^2 + G_6y_6{}^2 + G_7y_7{}^2\right)$$

Then using linear system (10.18) in $\frac{dU_2}{dt}$ we get

$$
\begin{aligned}
\frac{dU_2}{dt} =& G_1 \left[-(1-a)\beta_1 \left(\hat{S}_1 + y_1 \right) y_1 y_7 - (1-a)\beta_1 \hat{P} \, y_1{}^2 - \mu y_1{}^2 + \vartheta_1 y_1 y_3 \right] \\
&+ G_2 \left[(1-a)\beta_1 \left(\hat{S}_1 + y_1 \right) y_2 y_7 + (1-a)\beta_1 \hat{P} \, y_1 y_2 - (\alpha_1 + \mu + \sigma_1) y_2{}^2 \right] \\
&+ G_3 \left[\alpha_1 y_2 y_3 - (\mu + \vartheta_1) y_3{}^2 \right] \\
&+ G_4 \left[-(1-a)\beta_2 \left(\hat{S}_2 + y_4 \right) y_4 y_7 - (1-a)\beta_2 \hat{P} y_4{}^2 - \mu y_4{}^2 + \vartheta_2 y_4 y_6 \right] \\
&+ G_5 \left[(1-a)\beta_2 \left(\hat{S}_2 + y_4 \right) y_5 y_7 + (1-a)\beta_2 \hat{P} y_4 y_5 - (\alpha_2 + \mu + \sigma_2) y_5{}^2 \right] \\
&+ G_6 \left[\alpha_2 y_5 y_6 - (\mu + \vartheta_2) y_6{}^2 \right] \\
&+ G_7 \left[\eta y_5 y_7 - \delta y_7{}^2 \right]
\end{aligned}
$$

Now using the inequality, $\pm 2\, x\, y \le x^2 + y^2$ on the R.H.S. of $\frac{dU_2}{dt}$, we get;

$$
\begin{aligned}
\frac{dU_2}{dt} \le& G_1 \left[(1-a)\beta_1 \frac{\Delta_1}{2\mu} \left(y_1{}^2 + y_7{}^2 \right) - (1-a)\beta_1 \hat{P} \, y_1{}^2 - \mu y_1{}^2 + \vartheta_1 y_1 y_3 \right] \\
&+ G_2 \left[(1-a)\beta_1 \frac{\Delta_1}{2\mu} \left(y_2{}^2 + y_7{}^2 \right) + (1-a)\beta_1 \hat{P} \, y_1 y_2 - (\alpha_1 + \mu + \sigma_1) y_2{}^2 \right] \\
&+ G_3 \left[\alpha_1 y_2 y_3 - (\mu + \vartheta_1) y_3{}^2 \right] \\
&+ G_4 \left[(1-a)\beta_2 \frac{\Delta_2}{2\mu} \left(y_4{}^2 + y_7{}^2 \right) - (1-a)\beta_2 \hat{P} y_4{}^2 - \mu y_4{}^2 + \vartheta_2 y_4 y_6 \right] \\
&+ G_5 \left[(1-a)\beta_2 \frac{\Delta_2}{2\mu} \left(y_5{}^2 + y_7{}^2 \right) + (1-a)\beta_2 \hat{P} y_4 y_5 - (\alpha_2 + \mu + \sigma_2) y_5{}^2 \right] \\
&+ G_6 \left[\alpha_2 y_5 y_6 - (\mu + \vartheta_2) y_6{}^2 \right] \\
&+ G_7 \left[\eta y_5 y_7 - \delta y_7{}^2 \right]
\end{aligned}
$$

On arranging the above inequality we get;

$$
\begin{aligned}
\frac{dU_2}{dt} \le& - \left[\left((1-a)\beta_1 \hat{P} + \mu - (1-a)\beta_1 \frac{\Delta_1}{2\mu} \right) G_1 y_1{}^2 - \vartheta_1 G_1 y_1 y_3 \right. \\
&+ \left((\alpha_1 + \mu + \sigma_1) - (1-a)\beta_1 \frac{\Delta_1}{2\mu} \right) G_2 y_2{}^2 - (1-a)\beta_1 \hat{P} G_2 y_1 y_2 \\
&- \alpha_1 G_3 y_2 y_3 + (\mu + \vartheta_1) G_3 y_3{}^2 + \left((1-a)\beta_2 \hat{P} + \mu - (1-a)\beta_2 \frac{\Delta_2}{2\mu} \right) \\
&G_4 y_4{}^2 - \vartheta_2 G_4 y_4 y_6 + \left((\alpha_2 + \mu + \sigma_2) - (1-a)\beta_2 \frac{\Delta_2}{2\mu} \right) G_5 y_5{}^2 \\
&\left. - (1-a)\beta_2 \hat{P} G_5 y_4 y_5 - \alpha_2 G_6 y_5 y_6 + (\mu + \vartheta_2) G_6 y_6{}^2 - \eta G_7 y_5 y_7 \right.
\end{aligned}
$$

$$+ \left(\delta G_7 - (1-a)\,\beta_1 \frac{\Delta_1}{2\mu}\,(G_1 + G_2) - (1-a)\,\beta_2 \frac{\Delta_2}{2\mu}\,(G_4 + G_5) \right) y_7{}^2 \Bigg]$$

This implies;

$$\frac{dU_2}{dt} \le - \big[h_{11}\,y_1{}^2 - h_{13}y_1 y_3 + h_{22}y_2{}^2 - h_{12}\,y_1 y_2 - h_{23}y_2 y_3$$
$$+ h_{33}y_3{}^2 + h_{44}y_4{}^2 - h_{46}y_4 y_6 + h_{55}y_5{}^2 - h_{45}y_4 y_5$$
$$- h_{56}y_5 y_6 + h_{66}y_6{}^2 - h_{57}y_5 y_7 + h_{77}y_7{}^2 \big]$$

Where

$$h_{11} = \left((1-a)\,\beta_1 \hat{P} + \mu - (1-a)\,\beta_1 \frac{\Delta_1}{2\mu} \right) G_1,$$

$$h_{22} = \left((\alpha_1 + \mu + \sigma_1) - (1-a)\,\beta_1 \frac{\Delta_1}{2\mu} \right) G_2,$$

$$h_{33} = (\mu + \vartheta_1)\, G_3, h_{44} \left((1-a)\,\beta_2 \hat{P} + \mu - (1-a)\,\beta_2 \frac{\Delta_2}{2\mu} \right) G_4, h_{13} = \vartheta_1 G_1,$$

$$h_{55} = \left((\alpha_2 + \mu + \sigma_2) - (1-a)\,\beta_2 \frac{\Delta_2}{2\mu} \right) G_5$$

$$h_{66} = (\mu + \vartheta_2)\, G_6, h_{12} = (1-a)\,\beta_1 \hat{P}\, F G_2,$$

$$h_{77} = \delta G_7 - (1-a)\,\beta_1 \frac{\Delta_1}{2\mu}\,(G_1 + G_2) - (1-a)\,\beta_2 \frac{\Delta_2}{2\mu}\,(G_4 + G_5),$$

$$h_{23} = \alpha_1 G_3, h_{46} = \vartheta_2 G_4, h_{45} = (1-a)\,\beta_2 \hat{P} G_5, h_{56} = \alpha_2 G_6, h_{57} = \eta G_7.$$

again rearranging the inequality we have;

$$\frac{dU_2}{dt} \le - \Bigg[\left(\frac{h_{11}}{2}y_1{}^2 - h_{13}y_1 y_3 + \frac{h_{33}}{2}y_3{}^2 \right) . + \left(\frac{h_{11}}{2}y_1{}^2 - h_{12}\,y_1 y_2 + \frac{h_{22}}{2}y_2{}^2 \right)$$
$$+ \left(\frac{h_{22}}{2}y_2{}^2 - h_{23}y_2 y_3 + \frac{h_{33}}{2}y_3{}^2 \right) + \left(\frac{h_{44}}{2}y_4{}^2 - h_{46}y_4 y_6 + \frac{h_{66}}{2}y_6{}^2 \right)$$
$$+ \left(\frac{h_{44}}{2}y_4{}^2 - h_{45}y_4 y_5 + \frac{h_{55}}{3}y_5{}^2 \right) + \left(\frac{h_{55}}{3}y_5{}^2 - h_{56}y_5 y_6 + \frac{h_{66}}{2}y_6{}^2 \right)$$
$$+ \left(\frac{h_{55}}{3}y_5{}^2 - h_{57}y_5 y_7 + h_{77}y_7{}^2 \right) \Bigg] \tag{10.20}$$

Implementing the Sylvester criteria on the R.H.S. of inequality (10.20); under below mentioned conditions $\frac{dU_2}{dt}$ is negative definite;

$$\left((1-a)\,\beta_1 \hat{P} + \mu - (1-a)\,\beta_1 \frac{\Delta_1}{2\mu} \right) (\mu + \vartheta_1)\, G_3 > \vartheta_1{}^2 G_1,$$

$$\left((1-a)\,\beta_1\hat{P} + \mu - (1-a)\,\beta_1\frac{\Delta_1}{2\mu} \right) \left((\alpha_1 + \mu + \sigma_1) - (1-a)\,\beta_1\frac{\Delta_1}{2\mu} \right) G_1 >$$

$$\left((1-a)\,\beta_1\hat{P} \right)^2 G_2, \left((\alpha_1 + \mu + \sigma_1) - (1-a)\,\beta_1\frac{\Delta_1}{2\mu} \right) (\mu + \vartheta_1)\, G_2 > \alpha_1{}^2 G_3,$$

$$\left((1-a)\,\beta_2\hat{P} + \mu - (1-a)\,\beta_2\frac{\Delta_2}{2\mu} \right) (\mu + \vartheta_2)\, G_6 > \vartheta_2{}^2 G_4,$$

$$\frac{2}{3}\left((1-a)\,\beta_2\hat{P} + \mu - (1-a)\,\beta_2\frac{\Delta_2}{2\mu} \right) \left((\alpha_2 + \mu + \sigma_2) - (1-a)\,\beta_2\frac{\Delta_2}{2\mu} \right) G_4 >$$

$$\left((1-a)\,\beta_2\hat{P} \right)^2 G_5, \frac{2}{3}\left((\alpha_2 + \mu + \sigma_2) - (1-a)\,\beta_2\frac{\Delta_2}{2\mu} \right) (\mu + \vartheta_2)\, G_5 > \alpha_2{}^2 G_6,$$

$$\left\{ \frac{4}{3}\left(\delta G_7 - (1-a)\,\beta_1\frac{\Delta_1}{2\mu}(G_1 + G_2) - (1-a)\,\beta_2\frac{\Delta_2}{2\mu}(G_4 + G_5) \right) \right.$$

$$\left. \left((\alpha_2 + \mu + \sigma_2) - (1-a)\,\beta_2\frac{\Delta_2}{2\mu} \right) G_5 \right\} > (\eta G_7)^2 \qquad (10.21)$$

Now taking $G_1 = G_2 = G_4 = G_5 = 1$ and choosing G_3, G_6, G_7

$$\frac{2\mu\vartheta_1{}^2}{\left(2\mu\left((1-a)\,\beta_1\hat{P} + \mu \right) - (1-a)\,\beta_1\Delta_1 \right)(\mu + \vartheta_1)} < G_3$$

$$< \frac{(2\mu\,(\alpha_1 + \mu + \sigma_1) - (1-a)\,\beta_1\Delta_1)(\mu + \vartheta_1)}{2\mu\alpha_1{}^2},$$

$$\frac{2\mu\vartheta_2{}^2}{\left(2\mu\left((1-a)\,\beta_2\hat{P} + \mu \right) - (1-a)\,\beta_2\Delta_2 \right)(\mu + \vartheta_2)} < G_6$$

$$< \frac{(2\mu\,(\alpha_2 + \mu + \sigma_2) - (1-a)\,\beta_2\Delta_2)(\mu + \vartheta_2)}{3\mu\alpha_2{}^2},$$

$$G_7 > \frac{(1-a)\,(\beta_1\Delta_1 + \beta_2\Delta_2)}{\mu\delta}$$

Implementing Sylvester's criteria, under below mentioned conditions $\frac{dU_2}{dt}$ is negative definite;

$$2\mu\left((1-a)\,\beta_1\hat{P} + \mu \right) > (1-a)\,\beta_1\Delta_1,$$

$$2\mu\,(\alpha_1 + \mu + \sigma_1) > (1-a)\,\beta_1\Delta_1,$$

$$2\mu\left((1-a)\,\beta_2\hat{P} + \mu \right) > (1-a)\,\beta_2\Delta_2$$

$$and\, 2\mu\,(\alpha_2 + \mu + \sigma_2) > (1-a)\,\beta_2\Delta_2.$$

Hence from the Lyapnov's therem it can be found that endemic equilibrium point E_1 is non linearly stable under above mentioned conditions, otherwise unstable.

10.7 Sensitivity Analysis of Basic Reproduction Number and State Variables

To evaluate Sensitivity indices (significance of parameter and state variable on transmission and control of infection) the normalized forward sensitivity index method is used which is given by following formula;

$$i_x^{X_i} = \frac{\partial X_i}{\partial x_i} \cdot \frac{x_i}{X_i}$$

The parameter values and sensitivity indices of R_0 w. r. t. various parameters are stated in Table 10.1.

Positive a sign of sensitivity indices represents that the parameter is positively associated with state variable and basic reproduction number. Similarly, the negative sign of sensitivity indices represents that the parameter is negatively associated with the state variable and basic reproduction number. From Tables 10.1 and 10.2, we can conclude that as $\Delta_1, \Delta_2, \beta_1, \beta_2$ and η increase the influence of disease i.e. infective population also increases. While as $\mu, \alpha_1, \alpha_2, \sigma_1, \sigma_2, \delta$ and a increase the disease will be washed out from the population.

10.8 Optimal Control Problem

We propose an optimal control problem, which affects the transmission of droplet infection described in the model system (10.1) to (10.7). Here, a is a control strategy for the decline in the transmission rate of disease by using a mask used by the

Table 10.1   Sensitivity indices of R_0 with respect to parameter x_i

Parameters x_i	Parameter values	Sensitivity indices for R_0 $R_0 = 1.2987$
Δ_1	25	0.00
Δ_2	05	+1.00
β_1	0.015	0.00
β_2	0.02	+1.00
μ	0.4	-1.26
α_1	0.24	0.00
α_2	0.24	-0.16
σ_1	0.8	0.00
σ_2	0.9	-0.58
ϑ_1	0.8	0.00
ϑ_2	0.8	0.00
δ	0.3	-1.0
a	0.6	-1.5
η	06	+1.00

Table 10.2   Sensitivity indices of state variables with respect to parameter

Parameters x_i	Parameter values	$i_{x_i}^{S_1}$	$i_{x_i}^{I_1}$	$i_{x_i}^{R_1}$	$i_{x_i}^{S_2}$	$i_{x_i}^{I_2}$	$i_{x_i}^{R_2}$	$i_{x_i}^{P}$
		$\hat{S}_1$=51.14	$\hat{I}_1$=3.55	$\hat{R}_1$=0.71	$\hat{S}_2$=9.63	$\hat{I}_2$=0.83	$\hat{R}_2$=0.17	$\hat{P}$=16.67
Δ_1	25	+1.00	+1.00	+1.00	0.00	0.00	0.00	0.00
Δ_2	05	-0.79	+3.56	+3.56	0.00	+4.35	+4.35	+4.35
β_1	0.015	-0.18	+0..82	+0.82	0.00	0.00	0.00	0.00
β_2	0.02	-0.61	+2.74	+2.74	- 481.25	+3.35	+3.35	+3.35
μ	0.4	-0.01	-4.83	-5.16	+6.25	-4.55	-4.88	-4.55
α_1	0.24	+0.02	-0.15	+0.85	0.00	0.00	0.00	0.00
α_2	0.24	+0.11	-0.47	-0.47	+6.25	-0.58	+0.42	-0.58
σ_1	0.80	-0.013	-0.57	-0.57	0.00	0.00	0.00	0.00
σ_2	0.90	+0.47	-2.13	-2.13	+6.25	-2.61	-2.61	-2.61
ϑ_1	0.80	+0.01	+0.01	-0.66	0.00	0.00	0.00	0.00
ϑ_2	0.80	-0.01	+0.03	+0.03	0.00	+0.04	-0.63	+0.04
δ	0.30	+0.79	-3.56	-3.56	+32.08	-3.35	-3.35	-4.35
a	0.60	+1.19	-5.34	-5.34	+24.06	-5.02	-5.02	-5.02
η	06	-0.79	+3.56	+3.56	-1.60	+3.35	+3.35	4.35

population, u_1 is control strategies for reduction in the growth rate of pathogens by good hygiene and sanitation and u_2 is control strategies for increment in the death rate of pathogens by using inhalers and medicines. All three control strategies are used for minimizing the density of infected human beings, pathogens, and the expenditure related to controls a, u_1 and u_2. Now, we modify the model system (10.1) as follows;

$$\frac{dS_1}{dt} = \Delta_1 - (1-a)\beta_1 S_1 P - \mu S_1 + \vartheta_1 R_1$$

$$\frac{dI_1}{dt} = (1-a)\beta_1 S_1 P - (\alpha_1 + \mu + \sigma_1) I_1$$

$$\frac{dR_1}{dt} = \alpha_1 I_1 - (\mu + \vartheta_1) R_1$$

$$\frac{dS_2}{dt} = \Delta_2 - (1-a)\beta_2 S_2 P - \mu S_2 + \vartheta_2 R_2$$

$$\frac{dI_2}{dt} = (1-a)\beta_2 S_2 P - (\alpha_2 + \mu + \sigma_2) I_2$$

$$\frac{dR_2}{dt} = \alpha_2 I_2 - (\mu + \vartheta_2) R_2$$

$$\frac{dP}{dt} = (1-u_1)\eta I_2 - (\delta + u_2) P$$

We are using Pontryagin's Minimum Principle to study the optimal values of parameters. Further, suppose the Objective Functional J that optimizes the density of infected population, pathogen, and the expenditure related to control profiles as follows;

$$J\left(a\left(t\right),u_1\left(t\right),u_2\left(t\right)\right)=\int_o^{t_f}\left[A_1I_1\left(t\right)+A_2I_2\left(t\right)+A_3P\left(t\right)+\frac{A_4}{2}\left(a\right)^2+\frac{A_5}{2}\right.$$
$$\left.\left(u_1\right)^2+\frac{A_6}{2}\left(u_2\right)^2\right]dt \qquad (10.22)$$

The parameters $A_1>0$, $A_2>0$, $A_3>0$, $A_4>0, A_5>0$ and $A_6>0$ are dimensionless weight constants. Now suppose Lagrangian function L defined by;

$$L\left(I_1,I_2,P,a,u_1,u_2\right)=A_1I_1\left(t\right)+A_2I_2\left(t\right)+A_3P\left(t\right)+\frac{A_4}{2}\left(a\right)^2+\frac{A_5}{2}\left(u_1\right)^2$$
$$+\frac{A_6}{2}\left(u_2\right)^2$$

We get an optimal control parameter a^*, $u_1{}^*$ and $u_2{}^*$ as follows;

$$J\left(a^*,u_1{}^*,u_2{}^*\right)=\min_{a,\,u_1,u_2}\left[J\left(a,\,u_1,u_2\right)|a,\,u_1,u_2\in U\right]$$

Where $U=\{a,\,u_1,u_2|\,0\le a,\,u_1\,,\,u_2\le1\text{ and }t\in[0,t_f]\}$

Again, we consider Hamiltonian function as follows;

$$H=A_1I_1\left(t\right)+A_2I_2\left(t\right)+A_3P\left(t\right)+\frac{A_4}{2}\left(a\right)^2+\frac{A_5}{2}\left(u_1\right)^2+\frac{A_6}{2}\left(u_2\right)^2$$
$$+\lambda_1\left[\Delta_1-\left(1-a\right)\beta_1S_1P-\mu S_1+\vartheta_1R_1\right]+\lambda_2\left[\left(1-a\right)\beta_1S_1P\right.$$
$$-\left.\left(\alpha_1+\mu+\sigma_1\right)I_1\right]+\lambda_3\left[\alpha_1I_1-\left(\mu+\vartheta_1\right)R_1\right]$$
$$+\lambda_4\left[\Delta_2-\left(1-a\right)\beta_2S_2P-\mu S_2+\vartheta_2R_2\right]+\lambda_5\left[\left(1-a\right)\beta_2S_2P\right.$$
$$-\left.\left(\alpha_2+\mu+\sigma_2\right)I_2\right]+\lambda_6\left[\alpha_2I_2-\left(\mu+\vartheta_2\right)R_2\right]$$
$$+\lambda_7\left[\left(1-u_1\right)\eta I_2-\left(\delta+u_2\right)P\right]$$

where $\lambda_1,\lambda_2,\lambda_3,\lambda_4,\lambda_5,\lambda_6,\lambda_7$ are the adjoint variables that can be evaluated as follows;

$$\frac{d\lambda_1}{dt}=-\frac{\partial H}{\partial S_1},\frac{d\lambda_2}{dt}=-\frac{\partial H}{\partial I_1},\frac{d\lambda_3}{dt}=-\frac{\partial H}{\partial R_1},\frac{d\lambda_4}{dt}=-\frac{\partial H}{\partial S_2},\frac{d\lambda_5}{dt}=-\frac{\partial H}{\partial I_2},$$
$$\frac{d\lambda_6}{dt}=-\frac{\partial H}{\partial R_2},\frac{d\lambda_7}{dt}=-\frac{\partial H}{\partial P}.$$

$$\frac{d\lambda_1}{dt} = \lambda_1 \left[(1-a)\beta_1 P + \mu \right] - \lambda_2 \left[(1-a)\beta_1 P \right],$$

$$\frac{d\lambda_2}{dt} = -A_1 + \lambda_2 \left[\alpha_1 + \mu + \sigma_1 \right] - \lambda_3 \left[\alpha_1 \right],$$

$$\frac{d\lambda_3}{dt} = -\lambda_1 \left[\vartheta_1 \right] + \lambda_3 \left[\mu + \vartheta_1 \right],$$

$$\frac{d\lambda_4}{dt} = \lambda_4 \left[(1-a)\beta_2 P + \mu \right] - \lambda_5 \left[(1-a)\beta_2 P \right],$$

$$\frac{d\lambda_5}{dt} = -A_2 + \lambda_5 \left[\alpha_2 + \mu + \sigma_2 \right] - \lambda_6 \left[\alpha_2 \right] - \lambda_7 \left[(1-u_1)\eta \right],$$

$$\frac{d\lambda_6}{dt} = -\lambda_4 \left[\vartheta_2 \right] + \lambda_6 \left[\mu + \vartheta_2 \right],$$

$$\frac{d\lambda_7}{dt} = -A_3 + \lambda_1 \left[(1-a)\beta_1 S_1 \right] - \lambda_2 \left[(1-a)\beta_1 S_1 \right] + \lambda_4 \left[(1-a)\beta_2 S_2 \right]$$
$$- \lambda_5 \left[(1-a)\beta_2 S_2 \right] + \lambda_7 \left[\delta + u_2 \right]$$

with transversality conditions,

$$\lambda_1(T) = \lambda_2(T) = \lambda_3(T) = \lambda_4(T) = \lambda_5(T) = \lambda_6(T) = \lambda_7(T) = 0$$

The optimal controls can be noted by the following terms-
$$a^*(t) = \max\left\{ 0, \min\left(\hat{a}(t), 1 \right) \right\}, \quad u_1^*(t) = \max\left\{ 0, \min\left(\hat{u}_1(t), 1 \right) \right\},$$
$$u_2^*(t) = \max\left\{ 0, \min\left(\hat{u}_2(t), 1 \right) \right\}.$$ Further, greatest values and lowest values of the controls lies between zero and one respectively, on this basis we can conclude the following:

$$a^* = \begin{cases} 0 & \text{if } \hat{a} \leq 0, \\ \hat{a} & \text{if } 0 < \hat{a} < 1, \\ 1 & \text{if } \hat{a} \geq 1, \end{cases} \qquad u_1^* = \begin{cases} 0 & \text{if } \hat{u}_1 \leq 0, \\ \hat{u}_1 & \text{if } 0 < \hat{u}_1 < 1, \\ 1 & \text{if } \hat{u}_1 \geq 1, \end{cases} .$$

$$u_2^* = \begin{cases} 0 & \text{if } \hat{u}_2 \leq 0, \\ \hat{u}_2 & \text{if } 0 < \hat{u}_2 < 1, \\ 1 & \text{if } \hat{u}_2 \geq 1, \end{cases}$$

The zero indicates the inefficiency of a control strategy. Whereas one highlights the 100% effectiveness of the control strategy, in this case, there will be the highest decline in the infected population and pathogens. It is also pointed out that a^* and u_2^* are most effective control parameters for reducing infected population and pathogens, respectively (See Figure 10.2.

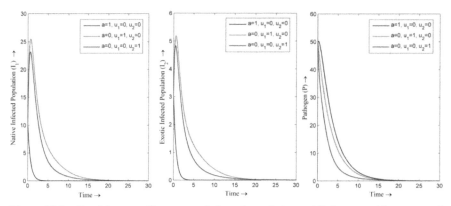

Figure 10.2   Graph between Time versus Infected population and Pathogens with one control out of three controls.

Again, differentiating H with respect to a, u_1 and u_2, we have;

$$\frac{\partial H}{\partial a} = A_4\,(a) + \lambda_1\,[\beta_1 S_1 P] - \lambda_2\,[\beta_1 S_1 P] + \lambda_4\,[\beta_2 S_2 P] - \lambda_5\,[\beta_2 S_2 P]$$

$$\frac{\partial H}{\partial u_1} = A_5\,(u_1) - \lambda_7\,[\eta I_2]$$

$$\frac{\partial H}{\partial u_2} = A_6\,(u_2) - \lambda_7\,[P]$$

The control classification, $\hat{a}$, $\hat{u}_1$ and $\hat{u}_2$ of the optimal controls a^*, $u_1{}^*$ and $u_2{}^*$ are attained by substituting $\frac{\partial H}{\partial a} = \frac{\partial H}{\partial u_1} = \frac{\partial H}{\partial u_2} = 0$ which gives,

$$\hat{a} = \frac{[(\lambda_2 - \lambda_1)\,\beta_1 S_1 + (\lambda_5 - \lambda_4)\,\beta_2 S_2]\,P}{A_4}\quad \hat{u}_1 = \frac{\lambda_7 \eta I_2}{A_5}\quad \hat{u}_2 = \frac{\lambda_7 P}{A_6}$$

10.9 Numerical Simulation

In this section, numerical simulation is performed for model system (10.1) to (10.7) using MATLAB and MATHEMATICA with the set of parametric values. Basic reproduction number (R_0) is influenced by parameters $(a, \beta_2, \delta, \eta)$ further, for the set of parameter values

$\Delta_1 = 25$, $\Delta_2 = 5$, $\beta_1 = 0.015$, $\alpha_1 = 0.24$, $\alpha_2 = 0.24$, $\sigma_1 = 0.8$, $\sigma_2 = 0.9$, $\vartheta_1 = 0.8$, $\vartheta_2 = 0.8$, $\mu = 0.4$ we have calculated basic reproductive number and equilibrium points.

Now, basic reproductive number and equilibrium points for various values of (β_2) are listed in Table 10.3.

Table 10.3 Basic Reproductive Number (R_0) and Equilibrium points for various values of (β_2)

Disease transmission rate (β_2)	Basic Reproductive Number (R_0)	Equilibrium points
0.015	0.9740	E_0 (62.5, 0, 0, 12.5, 0, 0, 0)
0.02	1.2987	E_1 (51.1364, 3.5511, 0.7102, 9.625, 0.8333, 0.1666, 16.6667)
0.03	1.9480	E_1 (42.511, 6.2465, 1.2493, 6.4166, 1.7632, 0.3526, 35.2657)
0.04	2.5974	E_1 (39.2045, 7.2798, 1.4559, 4.8125, 2.2282, 0.4456, 44.5652)

It is clear as β_2 increases then the disease free State leads to endemic state.

The basic reproductive number and equilibrium points for various values of (a) are listed in Table 10.4.

It is observed that as a increases then we arrive to disease free state from endemic state.

The basic reproductive number and equilibrium points for various values of (η) are listed in Table 10.5.

Table 10.4 Basic Reproductive Number (R_0) and Equilibrium points for various values of (a)

Mask efficiency (a)	Basic Reproductive Number (R_0)	Equilibrium points
0.2	2.5974	E_1 (28.5596, 10.6064, 2.1212, 4.8125, 2.2282, 0.4456, 44.5652)
0.4	1.9481	E_1 (36.6501, 8.0780, 1.6156, 6.4166, 1.7632, 0.3526, 35.2657)
0.6	1.2987	E_1 (51.1364, 3.5511, 0.7102, 9.625, 0.8333, 0.1666, 16.6667)
0.7	0.9740	E_0 (62.5, 0, 0, 12.5, 0, 0, 0)

Table 10.5 Basic Reproductive Number (R_0) and Equilibrium points for various values of (η)

Produce rate of pathogens (η)	Basic Reproductive Number (R_0)	Equilibrium points
4	0.8658	E_0 (62.5, 0, 0, 12.5, 0, 0, 0)
6	1.2987	E_1 (51.1364, 3.5511, 0.7102, 9.625, 0.8333, 0.1666, 16.6667)
7	1.5152	E_1 (45.1834, 5.4114, 1.0822, 8.25, 1.2319, 0.2463, 28.744)
8	1.7316	E_1 (40.4718, 6.8838, 1.3767, 7.2187, 1.5308, 0.3061, 40.8213)

Table 10.6 Basic Reproductive Number (R_0) and Equilibrium points for various values of (δ)

Death rate of pathogens (δ)	Basic Reproductive Number (R_0)	Equilibrium points
0.1	3.8961	E_1 (19.8124, 13.3399, 2.6679, 3.2083, 2.6932, 0.5386, 161.594)
0.2	1.9481	E_1 (36.6501, 8.0780, 1.6156, 6.4166, 1.7632, 0.3526, 52.8986)
0.3	1.2987	E_1 (51.1364, 3.5511, 0.7102, 9.625, 0.8333, 0.1666, 16.6667)
0.4	0.9740	E_0 (62.5, 0, 0, 12.5, 0, 0, 0)

It can be seen as η increases then influence of infection also increases.

The basic reproductive number and equilibrium points for various values of (δ) are listed in Table 10.6.

It is found as (δ)increases then influence of infection decrease.

10.10 Conclusions

In this chapter, a mathematical model has been suggested to analyze the impact of disease spread from exotic populations to native populations by droplets. It can be seen that $R_0 < 1$ represents the stability of disease-free equilibrium point, consequently, after some time disease will die out from the population. Similarly, $R_0 > 1$ represents the stability of the endemic equilibrium point, therefore disease will persist in a population Figure 10.3.

In this study by keeping the value of transmission rate of disease $(\beta_2 = 0.02)$ produce rate of pathogens $(\eta = 6)$ and death rate of pathogens $(\delta = 0.3)$ fixed, for small value of efficiency of mask(a), infected population and pathogens are higher or vice versa and also when mask efficiency (a) enhances then the value of basic reproductive number (R_0) declines Figures 10.4 and 10.9.

Again, keeping the value of efficiency of mask $(a = 0.6)$ produce rate of pathogens $(\eta = 6)$ and death rate of pathogens $(\delta = 0.3)$ fixed, for small value of transmission rate of disease (β_2), infected population and pathogens are lower or

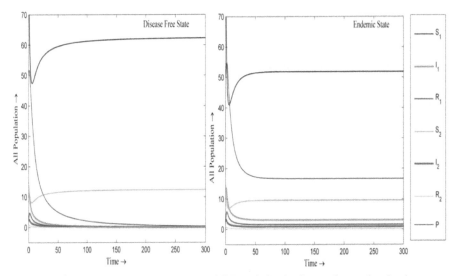

Figure 10.3   Graph between Time versus All Population in disease free and endemic states.

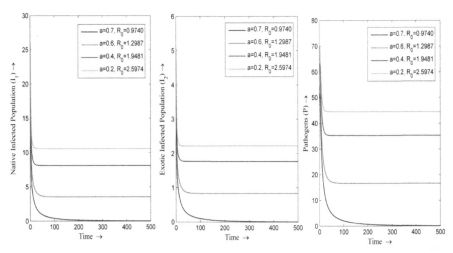

Figure 10.4 Graph between Time versus Infected population and Pathogens for various of (a).

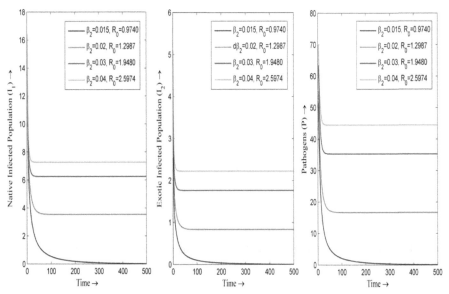

Figure 10.5 Graph between Time versus Infected population and Pathogens for various values of (β_2).

vice versa and also as transmission rate of disease (β_2) increases then the value of basic reproduction (R_0) increases. Figures 10.5 and 10.9. This is an indication that the transmission can be prevented by using appropriate measures.

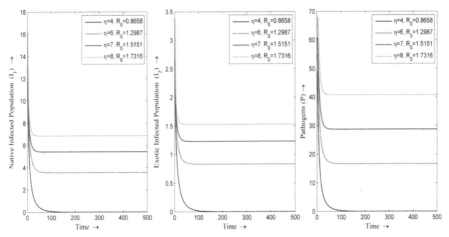

Figure 10.6   Graph between Time versus Infected population and Pathogens for various values of (η).

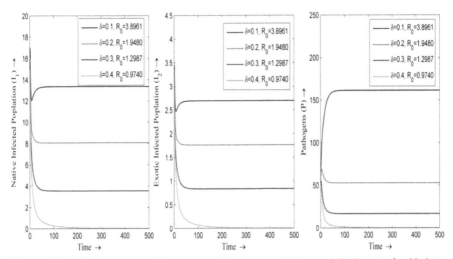

Figure 10.7   Graph between Time versus Infected population and Pathogens for Various values of (δ).

Moreover, keeping the value of mask efficiency $(a = 0.6)$ disease transmission rate $(\beta_2 = 0.02)$ and death rate of pathogens $(\delta = 0.3)$ fixed, for small value of production rate of pathogens (η), infected population and pathogens are lower or vice versa and also as produce rate of pathogens increases then the value of basic reproduction (R_0) increases. Figures 10.6, and 10.9.We can also conclude that pathogens like bacteria, viruses, fungi, etc. are responsible for disease transmission.

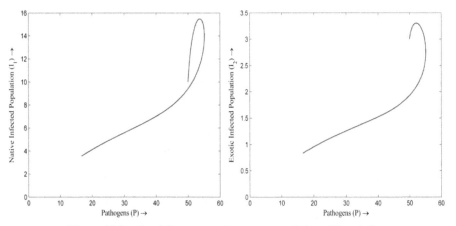

Figure 10.8   Graph between Pathogens versus Infected Population

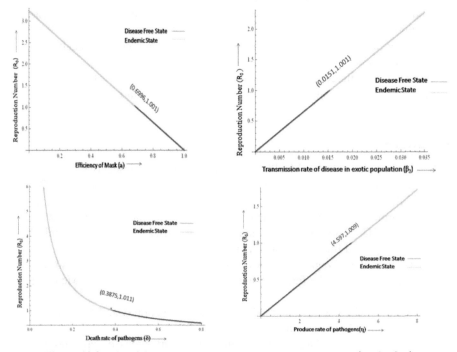

Figure 10.9   Graph between Basic Reproduction Number versus $(a, \beta_2, \delta, \eta)$.

Further, keeping the value of efficiency of mask efficiency ($a = 0.6$) disease transmission rate ($\beta_2 = 0.02$) produce and rate of pathogens ($\eta = 6$) fixed, for small value of death rate of pathogens (δ), infected population and pathogens are higher or vice versa and also as the death rate of pathogens increases then the value of basic reproduction (R_0) decreases. Figures 10.7, and 10.9 Bifurcation points are also heightened in Figure 10.9 for each case.

The Exotic infected population is the main cause of disease spread in the native population and get them infected. From Figure 10.9 it is obtained that as I_2 increases then I_1 also increases. In Figure 10.8 it is seen that as the number of pathogens increases the infected population also increases.

It is also seen that when control strategy (a) is taken into practice number of native and exotic infected is minimum and when the control strategy (u_2) is adopted a number of pathogens is minimum. Figure 10.2

On the basis of sensitivity indices, we can conclude the following suggestions;

1. Recruitment rate of native and exotic population Δ_1 and Δ_2 should be reduced to prevent the propagation of disease.
2. Transmission rate of infection β_1 and β_2 should be minimized by using a mask so that the disease cannot be transmitted.
3. Produce rate of pathogens η should be reduced by keeping good hygiene conditions and clean surroundings.
4. Recovery rate α_1 and α_2 should be boosted by maintaining good health and immune system.
5. Death rate of pathogens δ should be increased by availing proper treatment.
6. Efficient of mask a should be higher so that transmission of disease can be prohibited.

References

[1] B. D. Gushulak, D. W. MacPherson:"Globalization of infectious diseases: The impact of migration", (2004), Clinical Infectious Diseases, vol. 38, pp. 1742–1748.
[2] A. G. Carmichael: "Historical plague", (2014), Elsevier Indiana University.
[3] F. Castelli, G. Sulis:"Migration and infectious diseases", (2017), Clinical Microbiology and Infection, vol. 23, pp. 283-289.
[4] B. J. Coburn, B. G. Wagner, S. Blower:"Modeling influenza epidemics and pandemics: insights into the future of swine flu (H1N1)", (2009), BMC Medicine 7, Article no. 30.
[5] D. J. Daley, J. Gani:"Epidemic modelling: An introduction (Cambridge studies in mathematical biology", (2005).
[6] P. Van den Driessche:"Reproduction numbers of infectious disease model", (2017), Infectious Disease Modelling, vol. 2, pp. 288-303.
[7] J. M. Heffernan, R. J. Smith, L. M. Wahl:"Perspectives on the basic reproductive ratio", (2005), Journal of The Royal Society Interface, vol. 2, pp. 281–293.

[8] M. T. Li, Z. Jin, G. Q. Sun, J. Zhang:"Modeling direct and indirect disease transmission using multi-group model", (2017), Journal of Mathematical Analysis and Applications, vol. 446, pp. 1292-1309.

[9] O. P. Misra, P. Kushwah, C. S. Sikarwar: "Effect of resource based exotic goose species on native plant species competing with exotic grass: A model", (2013), Proceedings of the National Academy of Sciences, vol. 83(4), pp. 343–351.

[10] O. P. Misra, D. K. Mishra:"Modelling the effect of booster vaccination on the transmission dynamics of disease that spread by droplet infection", (2009), Nonlinear Analysis: Hybrid systems, vol. 3, pp. 657-665.

[11] K. Park:"Preventive and social medicine", (2002), M/S BanarsiDas Bhanot publishers. Jabalpur. India.

[12] K. Park:"Essentials of community health nursing", (2004), M/S BanarsiDas Bhanot publishers. Jabalpur. India.

[13] S. S. Pattnaik, K. M. Bakwad, B. S. Sohi, R. K. Ratho, S. Devi: "Swine influenza models based optimization (SIMBO)", (2013), Applied Soft Computing, vol. 13, pp. 628–653.

[14] A. J. Tatem, D. J. Rogers, S. I. Hay:"Global transport networks and infectious disease spread", (2006), Advances in Parasitology , vol. 62, pp. 293-343.

[15] C. Purushwani, H. Purushwani, R. Deolia: "Transmission dynamic of tuberculosis in two dissimilar groups through pathogens: A SIRS model", (2020), Malaya Journal of Matematik, vol. 8 (3), pp. 930-938.

[16] H. Purushwani, P. Sinha: "Impact of profession and surroundings on spread of swine flu: A mathematical study", (2019), Malaya Journal of Matematik, vol. 7, no. 2, pp. 276-286.

11

A Fractional Calculus Model to Depict the Calcium Diffusion for Neurodegenerative Disease

Hardik Joshi[1] and Brajesh Kumar Jha[2]

[1]Department of Mathematics, LJ Institute of Engineering and Technology,
LJ University, Ahmedabad-382210
[2]Department of Mathematics, School of Technology,
Pandit Deendayal Energy University, Gandhinagar-382007
E-mail: hardik.joshi8185@gmail.com; brajeshjha2881@gmail.com

Abstract

Neurodegenerative disease is widely spread across the world in recent times. There are several factors associated with that to reach up to these conditions such as oldest age, environmental effects, protein structure, misfolded gene, calcium dysfunction, and so on. Out of all listed factors, calcium dysfunction is the common target for various neurodegenerative diseases like Alzheimer's, Parkinson's, Huntington's, and many others. In the present chapter, we have developed the fractional calculus model to study the calcium diffusion in the cells. A several physiological parameters such as diffusion coefficient, advection, velocity, and amplitude source are taken into consideration correspond to the physical conditions of the cells. The analytic solution of the proposed model is derived by the use of the similarity and Laplace transform. The obtained results are shown the importance of the calcium diffusion in the neurodegenerative disease due to the global behavior of the fractional calculus.

Keywords: Calcium, Advection diffusion equation, Fractional calculus, Laplace transform, Neurodegenerative disease.

11.1 Introduction

Computational neuroscience or mathematical neuroscience is a specialized branch of neuroscience to understand the behavior of the nervous system by using a mathematical model. The mathematical model translates the biophysical characterization of the neuron into the mathematical formulation whose rigorous study reveals the insight of the original problem. Neuron is an assembly of the nerves that circulate the bidirectional information to the brain and our body. A disorder connected to the dysfunction of the nerves is called a neurodegenerative disease that widely covers Alzheimer's disease (AD), Parkinson's disease (PD), Huntington's disease (HD), and many others [1].

The first well known neurodegenerative disease is AD which start the form of dementia when an individual losses their memory. The progress of AD is generally categorized into three stages preclinical, mild, and severe due to its attribute and symptoms. The alteration in the amount of amyloid beta is responsible for degeneration and loss of brain functions. The symptoms of this can be visualized from the thinking and reasoning ability, change in the personality, apathy, depression, swing of mood, etc. There are several other phenomena associated with the cause of AD such as tau protein, change in genetics, oxidative stress, and disturbance in the calcium signaling [2–5].

After AD, Parkinson's is the second neurodegenerative disease whose at most symptoms are similar to AD. Thus in the preclinical stage, it is hard to identify the precise disease. Although the bradykinesia, rigidity in the muscle, tremors are communally associated with the PD and that cannot be visualized in the case of AD. Also, the clinical data suggest that the premature signs of AD and PD are normally seen in an individual after fifty and before sixty respectively. The loss in the memory is typically associated with AD but can also be seen in the case of PD. The alteration in the amount of dopamine neurons is responsible for the degeneration and loss of the brain functionality of the PD. Also, there are several other factors and phenomena associated with the early progression of PD [4–8].

Huntington's disease (HD) is an inherited rare neurodegenerative disease of nerve cells started near the age of 30. Due to the inheritance nature of HD, there is a chance of spread of chromosome 4 from parents to the child and this is a main responsible gene. The chromosome 4 damaged the protein sequence and made a more CAG protein sequence as against a requirement. That damages the entire nerve and progresses the early cause of the HD. As AD and PD there are so many other things that are also responsible for the disease but in the present study, we have explored the role of calcium diffusion in all these neurodegenerative diseases [4], [5], [9–11].

Calcium is required for everyone at every stage of life to perform the day to day activities in a normal way. The requirement of calcium varies from individual to individual corresponding to its day schedule. It manages the calcium signaling phenomena and is also involved in the major activities associated with the signaling process. The malfunction of this invites a variety of neurodegenerative diseases such as AD, PD, HD, and many others [4], [5], [12]. The precise mechanism of diffusion of calcium is studied by the researcher in the different cells like astrocyte cells [13–15], neuron cells [16–18], oocyte cells [19–21], myocytes cells [22–24], T lymphocyte cells [25–27], hepatocyte cells [28], pancreatic acinar cells [29]. A few attempts are registered to observe the malfunction of calcium diffusion in neurodegenerative diseases like AD [30–32], and PD [33–40]. Also, a little attempt is registered to observe the problems of the nervous system by using the fractional order method [40–42]. It is observed that the researchers widely used an analytical method with cartesian or a polar coordinate, numerical methods like finite difference method, finite element method, and finite volume method are used to solve mathematical models according to the structure or geometry of particular cells. Thus, available literature works suggest that there is little work done to study the calcium diffusion effect in neurodegenerative disease by using fractional calculus.

The later part of the study is as follows: First, we have defined a basic definition and its Laplace transformation. Then we have developed a mathematical model in the form of an advection diffusion equation later it is modified by using fractional calculus approaches. The results are obtained for various values of the fractional parameter and present and absent of advection flux and relate to neurodegenerative disease. At last, some conclusion derives from the present study.

11.2 Mathematical Preliminaries

Here, we have listed the basic definition of Riemann–Liouville, Caputo fractional derivative and its Laplace transform that can be used to solve the mathematical model [43–47].

11.2.1 The Riemann–Liouville fractional order integral for a function $y(t)$ is defined as

$$I_a^\alpha(y(t)) = \frac{1}{\Gamma(\alpha)} \int_a^t (t - \xi)^{\alpha-1} y(\xi)\, d\xi \tag{11.1}$$

Where $t > a$, and $R(\alpha) > 0$.

11.2.2 The Riemann–Liouville fractional order derivative for a function $y(t)$ of order α is defined as

$$_a^{RL}D_t^\alpha(y(t)) = \left(\frac{d}{dt}\right)^n (I_a^{n-\alpha}y(t)) \qquad (11.2)$$

Where $R(\alpha) > 0$.

11.2.3 The Caputo fractional order derivative for a function $y(t)$ of order α is defined as

$$_a^C D_t^\alpha(y(t)) = \begin{cases} \frac{1}{\Gamma(m-\alpha)} \int_a^t \frac{y^m(\xi)}{(t-\xi)^{\alpha+1-m}} \, d\xi, & m-1 < \alpha \le m \\ \frac{d^m}{dt^m} y(t), & \alpha = m \end{cases} \qquad (11.3)$$

Where $R(\alpha) > 0$ and $m \in N$.

11.2.4 The Laplace transform of function $y(t)$ with respect to time is defined as

$$L\{y(t)\} = \bar{y}(s) = \int_0^\infty e^{-st} y(t) \, dt \qquad (11.4)$$

Where $R(s) > 0$ and $t > 0$.

11.2.5 The inverse Laplace transform of function $\bar{y}(s)$ is defined as follows

$$L^{-1}\{\bar{y}(s)\} = y(t) = \frac{1}{2\pi i} \int_{\gamma-i\infty}^{\gamma+i\infty} e^{st} \bar{y}(s) \, ds \qquad (11.5)$$

Where $\gamma \in R$ is a some fixed number.

11.2.6 The Laplace transform of the Caputo fractional derivative is defined as

$$\int_0^\infty e^{-st} {_a^C D_t^\alpha}(y(t)) \, dt = s^\alpha y(s) - \sum_{j=0}^{n-1} s^{\alpha-j-1} y^j(0) \qquad (11.6)$$

Where $n-1 < \alpha < n$.

11.3 Mathematical Modelling of Advection Diffusion Equation

The rate of change of calcium depends on the calcium entering into the cells through a source of calcium flux, calcium out into the cytosol, and calcium diffuses into the cells for the production of other molecules.

The mathematical expression for this process is represented in the form of a differential equation as [48]

$$\frac{\partial u(x,t)}{\partial t} - \frac{\partial f(x,t)}{\partial x} = J(x,t,u) \tag{11.7}$$

Here $u(x,t)$ represents the concentration of calcium at position x and time t, $f(x,t)$ represents the rate of calcium flow, and $J(x,t,u)$ represents the final rate of the production and destruction of calcium per unit volume in the cells.

Here we have considered advection flux along with the diffusive flux so the total flux term is evaluated from the given differential equation

$$f(x,t) = v \cdot u(x,t) - D\frac{\partial u(x,t)}{\partial x} \tag{11.8}$$

Where D is the diffusion constant and v is the velocity.

According to the Fick" law the concentration of calcium is flow from the higher to lower concentration level in the cells. By using this fact, the equation (11.7) is rewritten in the form of an advection diffusion equation (ADE) as

$$\frac{\partial u(x,t)}{\partial t} - \frac{\partial(v \cdot u(x,t))}{\partial x} - D\frac{\partial u(x,t)}{\partial x} = J(x,t,u) \tag{11.9}$$

The present study is deal with the homogeneous ADE so the one-dimensional calcium diffusion model becomes as

$$\frac{\partial u(x,t)}{\partial t} - \frac{\partial(v \cdot u(x,t))}{\partial x} = D\frac{\partial^2 u(x,t)}{\partial x^2} \tag{11.10}$$

The initial and boundary conditions are taken corresponds to the physiology of cells and are given as follows

$$\lim_{x \to 0} \left(D\frac{\partial u(x,t)}{\partial x} \right) = \sigma, \quad t > 0 \tag{11.11}$$

$$\lim_{x \to 0} u(x,t) = 0, \quad t \geq 0 \tag{11.12}$$

$$u(x,0) = u_0, \quad 0 \leq x < \infty \tag{11.13}$$

The calcium flow through the cytosol to calcium reservoir entities and vice versa follows the Gaussian distribution all over the cells. It is assumed that the flow is begun from the resting of the cells; at the initial point of the source. Thus, it is an inviscid case and we assigned a new coordinate system to flow across the cells as

$$\xi = x - (x_0 + vt) \tag{11.14}$$

$$\tau = t \tag{11.15}$$

Where ξ and τ are the new spatial and time coordinates respectively.

The new coordinates system is applied in the homogeneous ADE and by incorporating the chain rule, we get

$$\frac{\partial u}{\partial \tau}\frac{\partial \tau}{\partial t} + \frac{\partial u}{\partial \xi}\frac{\partial \xi}{\partial t} + v\left(\frac{\partial u}{\partial \tau}\frac{\partial \tau}{\partial x} + \frac{\partial u}{\partial \xi}\frac{\partial \xi}{\partial x}\right)$$
$$= D\left(\frac{\partial}{\partial \tau}\frac{\partial \tau}{\partial x} + \frac{\partial}{\partial \xi}\frac{\partial \xi}{\partial x}\right)\left(\frac{\partial u}{\partial \tau}\frac{\partial \tau}{\partial x} + \frac{\partial u}{\partial \xi}\frac{\partial \xi}{\partial x}\right) \tag{11.16}$$

Further simplification of equation (11.16) gives us

$$\frac{\partial u(\xi,\tau)}{\partial \tau} = D\frac{\partial^2 u(\xi,\tau)}{\partial \xi^2} \tag{11.17}$$

The initial and boundary conditions are as follows

$$\lim_{\xi \to 0}\left(D\frac{\partial u(\xi,\tau)}{\partial \xi}\right) = \sigma, \quad \tau > 0 \tag{11.18}$$

$$\lim_{\xi \to \infty} u(\xi,\tau) = u_\infty, \quad \tau \geq 0 \tag{11.19}$$

$$u(\xi,0) = u_\infty, \quad 0 \leq \xi < \infty \tag{11.20}$$

The advection diffusion model (11.17)–(11.20) is used to obtain the calcium estimation in the cells. But the classical behavior of the ADE is not include the memory of cells to obtain the estimation of calcium. Thus, a fractional order mathematical model is developed to remove this disadvantage.

11.3.1 A Fractional Calculus Model of Advection Diffusion Equation

A fractional calculus model of calcium diffusion for neurodegenerative disease is obtained by replacing the classical derivative with the Caputo derivative in the equation (11.17) then we have

$$\frac{\partial^\alpha u(\xi,\tau)}{\partial \tau^\alpha} = D\frac{\partial^2 u(\xi,\tau)}{\partial \xi^2} \tag{11.21}$$

Where $0 < \alpha \leq 1$ and $\xi, \tau \geq 0$.

The initial and boundary conditions are as follows

$$\lim_{\xi \to 0} \left(D \frac{\partial u(\xi, \tau)}{\partial \xi} \right) = \sigma, \quad \tau > 0 \tag{11.22}$$

$$\lim_{\xi \to \infty} u(\xi, \tau) = u_\infty, \quad \tau \geq 0 \tag{11.23}$$

$$u(\xi, 0) = u_\infty, \quad 0 \leq \xi < \infty \tag{11.24}$$

Here σ represents source amplitude of calcium, and u_∞ represents the resting concentration of calcium. The classical ADE model is recovered by considering the value of $\alpha = 1$ in the equation (11.21).

By applying the Laplace transform on both the side of the equation (11.21) we have

$$s^\alpha \bar{u}(\xi, s) - s^{\alpha-1} u(\xi, 0) = D \frac{\partial^2 \bar{u}(\xi, s)}{\partial \xi^2} \tag{11.25}$$

At the same time, we apply the Laplace transform on the boundary conditions (11.22)–(11.23) then we get

$$\frac{\partial}{\partial \xi} \bar{u}(0, s) = \frac{\sigma}{D s^\alpha} \tag{11.26}$$

$$\lim_{\xi \to \infty} \bar{u}(\xi, s) = 0 \tag{11.27}$$

Thus the solution of equation (11.25) along with boundary and initial condition of the cell is becomes as

$$\bar{u}(\xi, s) = a \exp\left(\sqrt{\frac{s^\alpha}{D}} \xi \right) + b \exp\left(-\sqrt{\frac{s^\alpha}{D}} \xi \right) \tag{11.28}$$

Where exp is an exponent function and a, b are constants.

The values of the constant a and b are determined by using the equation (11.26) and (11.27) then the further simplification gives

$$a = 0, \quad b = \frac{\sigma}{\sqrt{D} s^{\alpha/2}} \tag{11.29}$$

Thus, by substituting the values of constants the equation (11.28) is rewritten as

$$\bar{u}(\xi, s) = \frac{\sigma}{\sqrt{D} s^{\alpha/2}} \exp\left(-\sqrt{\frac{s^\alpha}{D}} \xi \right) \tag{11.30}$$

Now by taking the inverse Laplace transform of equation (11.30) and using the equation (11.5) we get the following expression as [43–45]

$$u(\xi, \tau) = \frac{\sigma \tau^{\alpha/2-1}}{\sqrt{4D}} W \left[-\frac{\alpha}{2}, \frac{\alpha}{2}; -\frac{\xi}{\sqrt{D}} \tau^{-\alpha/2} \right] \qquad (11.31)$$

Finally, we convert the equation (11.31) into the original coordinate of the proposed model then we have

$$u(x, t) = \frac{\sigma t^{\alpha/2-1}}{\sqrt{4D}} W \left[-\frac{\alpha}{2}, \frac{\alpha}{2}; -\frac{x - vt}{\sqrt{D}} t^{-\alpha/2} \right] \qquad (11.32)$$

The equation (11.32) is used to obtain the estimation of calcium in the various positions and time in the cells to investigated the neuronal behaviour.

11.4 Results and Discussion

The graphical representation of the calcium profile is obtained by simulating the results derived in equation (11.32). The values of a numerical parameter to obtain the graphical representation are as follows [16], [19], [21], [49], [50]: source amplitude $\sigma = 1.6 \ \mu M^{-1}s^{-1}$, diffusion coefficients for calcium $D = 200 - 300 \ \mu m^2/s$, advection flux or velocity $v = 0 - 20 \ \mu m/s$, x and t are the components of the space or distance and time respectively.

Figures 11.1 and 11.2 shows the calcium profile against time for the presence and absence of velocity for different value of α a fractional derivative. The velocity flux for the presence and absence of advection flux is considered here is 20 and 0 $\mu m/s$ respectively. It is observed that the calcium profile suddenly attains the peak near $t = 0.1 \ s$ and then decreased at the background level. The classical calcium profile that is for $\alpha = 1$ early attains the background level as compared to the fractional order profile. The fractional profile attains a low background level due to the non-local properties of the Caputo derivative. Also, it is observed that in the absence of advection flux the calcium profile early attains the background level as seen in Figure 11.2. As the Caputo derivative goes from 1 to 0.8 the calcium profile is decreased over the region and protects the cells from degeneration produced by neurodegenerative disease.

Figure 11.3 show the calcium profile against time for various amounts of velocity flux $v = 0 - 20 \ \mu m/s$ at $\alpha = 1$. The peak value for the different amount of velocity flux is almost the same but there is a difference in attaining the background level. In the absence or low value of velocity flux, the calcium

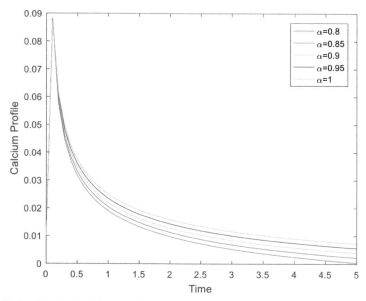

Figure 11.1 Graph of calcium profile against time for various fractional order derivative in presence of advection flux.

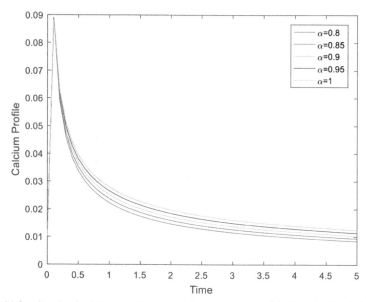

Figure 11.2 Graph of calcium profile against time for various fractional order derivative in absence of advection flux.

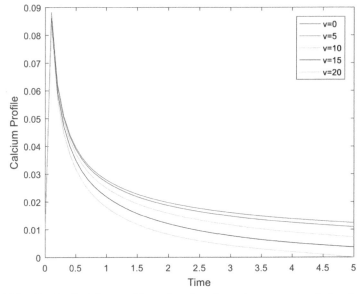

Figure 11.3   Graph of calcium profile against time for various amounts of advection flux at $\alpha = 1$.

profile early attains the background level. The higher amounts of velocity flux reduce the calcium profile and control against the neurodegenerative disease as high calcium is not suitable for cells for a long period.

Figures 11.4 and 11.5 shows the calcium profile against distance for presence and absence of velocity for different value of α. The velocity flux for the presence and absence of advection flux is the same as Figures 11.1 and 11.2 respectively. It is observed that the calcium profile is high at $x = 0\ \mu m$ and then slowly decreased to get the background level. The classical calcium profile that is for $\alpha = 1$ is high and early attains the background level due to the local behavior of the classical derivative. Also, it is observed that for the absence of advection flux the calcium profile is high as compared to the present as seen in Figure 11.4.

Figure 11.6 shows the calcium profile against distance for various amounts of velocity flux $v = 0 - 20\ \mu m/s$ at $\alpha = 1$. The peak value of calcium profile is high for the absence or low amount of velocity flux but the behavior is changed near $x = 14\ \mu m$ due to the non-local properties of Caputo derivative. Hence it is observed that the velocity flux also provides a significant impact on a calcium profile.

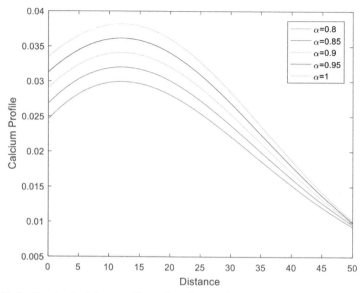

Figure 11.4 Graph of calcium profile against distance for various fractional order derivative in presence of advection flux.

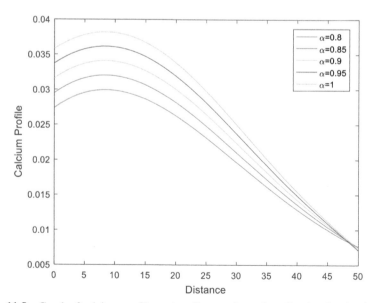

Figure 11.5 Graph of calcium profile against distance for various fractional order derivative in absence of advection flux.

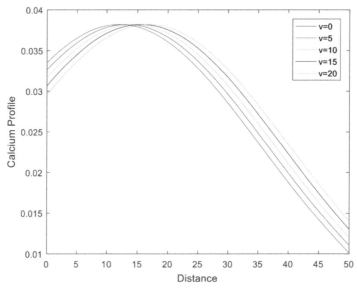

Figure 11.6   Graph of calcium profile against distance for various amounts of advection flux at $\alpha = 1$.

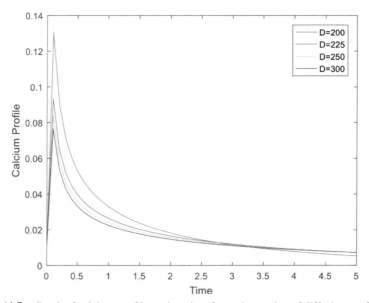

Figure 11.7   Graph of calcium profile against time for various value of diffusion coefficients at $\alpha = 1$.

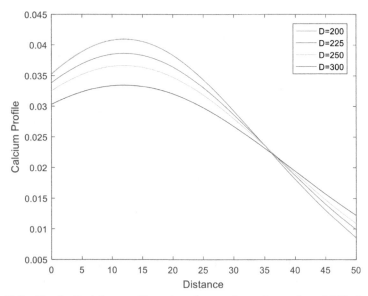

Figure 11.8 Graph of calcium profile against distance for various value of diffusion coefficients at $\alpha = 1$.

Figures 11.7 and 11.8 show the calcium profile against time and distance for various values of diffusion coefficients at $\alpha = 1$.It is observed that for $D = 200 \ \mu m^2/s$ the calcium profile attains a peak value. As the value of the diffusion coefficient increases from 200 to 300 $\mu m^2/s$ the calcium profile is decreased. Thus, it is observed that the diffusion coefficient also affects the calcium profile to attain a peak and background level.

Thus, the results clearly show that the presence and absence of advection flux, different amounts of velocity, diffusion coefficients, and fraction order derivative are directly related to calcium profile. The alteration in any of these parameters may culminate in neurodegenerative diseases such as AD, PD, HD, and many others.

11.5 Conclusion

In this chapter, we have presented the fractional calculus model for the advection-diffusion equation that arise in the calcium diffusion process of the neurodegenerative disease. We modified the classical advection diffusion equation into the Caputo advection diffusion equation to improve the accuracy of the model. We have framed the initial and boundary conditions corresponding to calcium diffusion phenomena. The present model is transformed into a new coordinate system to convert the advection diffusion equation into the Caputo diffusion equation. The solution is obtained in the generalized exponential function known as the Wright function by applying the Laplace integral transform technique. The calcium profiles are obtained graphically for the various order of the fractional derivative, presence, and absence of advection flux, velocity, and diffusion coefficients. It is observed that the velocity, diffusion coefficients and fractional order derivative plays a significant role due to the memory property of the Caputo model. Thus, it provides control to maintain the calcium profile in the neuron cells and as a consequence rescue the cell from neurodegenerative disease like AD, PD, and HD.

References

[1] L. R. Squire, D. Berg, F. E. Bloom, S. Du Lac, A. Ghosh, and N. C. Spitzer, "Fundamental Neuroscience: Fourth Edition," Fourth. Elsevier Inc., 2012.

[2] M. Goedert and M. G. Spillantini, "A century of Alzheimer's disease," Science, vol. 314, no. 5800. pp. 777–781, 03-Nov-2006.

[3] L. Mucke, "Neuroscience: Alzheimer's disease," Nature, vol. 461, no. 7266. Nature Publishing Group, pp. 895–897, 14-Oct-2009.

[4] M. P. Mattson, "Calcium and neurodegeneration," Aging Cell, vol. 6, no. 3, pp. 337–350, Jun. 2007.

[5] P. Marambaud, U. Dreses-Werringloer, and V. Vingtdeux, "Calcium signaling in neurodegeneration.," Mol. Neurodegener., vol. 4, p. 20, 2009.

[6] D. J. Surmeier, J. N. Guzman, and J. Sanchez-Padilla, "Calcium, cellular aging, and selective neuronal vulnerability in Parkinson's disease," Cell Calcium. 2010.

[7] A. H. Schapira, "Calcium dysregulation in Parkinson's disease," Brain A J. Neurol., vol. 136, no. 7, pp. 2015–2016, 2013.

[8] P. Rivero-Ríos, P. Gómez-Suaga, E. Fdez, and S. Hilfiker, "Upstream deregulation of calcium signaling in Parkinson's disease.," Front. Mol. Neurosci., vol. 7, no. June, p. 53, 2014.

[9] G. P. Bates et al., "Huntington disease," Nat. Rev. Dis. Prim. 2015 11, vol. 1, no. 1, pp. 1–21, Apr. 2015.

[10] I. Bezprozvanny and M. R. Hayden, "Deranged neuronal calcium signaling and Huntington disease," Biochem. Biophys. Res. Commun., vol. 322, no. 4, pp. 1310–1317, Oct. 2004.

[11] L. A. Raymond, "Striatal synaptic dysfunction and altered calcium regulation in Huntington disease," Biochem. Biophys. Res. Commun., vol. 483, no. 4, pp. 1051–1062, Feb. 2017.

[12] E. Carafoli and M. Brini, Calcium Signalling and Disease. Springer Science and Business Media, 2007.

[13] B. K. Jha, N. Adlakha, and M. N. Mehta, "Two-dimensional finite element model to study calcium distribution in astrocytes in presence of excess buffer," Int. J. Biomath., vol. 7, no. 3, pp. 1–11, 2014.

[14] B. K. Jha, A. Jha, and N. Adlakha, "Three-Dimensional finite element model to study calcium distribution in Astrocytes in presence of VGCC and excess Buffer," Differ. Equations Dyn. Syst., vol. 28, no. 3, pp. 603–616, Nov. 2020.

[15] B. K. Jha, N. Adlakha, and M. N. Mehta, "Two-Dimensional finite element model to study calcium distribution in Astrocytes in presence of VGCC and excess Buffer," Int. J. Model. Simulation, Sci. Comput., vol. 4, no. 2, 2013.

[16] A. Jha and N. Adlakha, "Two-dimensional finite element model to study unsteady state Ca2+ diffusion in neuron involving ER LEAK and SERCA," Int. J. Biomath., vol. 8, no. 1, pp. 1–14, 2015.

[17] A. Jha and N. Adlakha, "Analytical solution of two dimensional unsteady state problem of calcium diffusion in a neuron cell," J. Med. Imaging Heal. Informatics, vol. 4, no. 4, pp. 547–553, 2014.

[18] A. Jha, N. Adlakha, and B. K. Jha, "Finite element model to study effect of Na+ - Ca2+ exchangers and source geometry on calcium dynamics in a neuron cell," J. Mech. Med. Biol., vol. 16, no. 2, pp. 1–22, 2015.

[19] P. A. Naik and K. R. Pardasani, "Finite element model to study calcium distribution in oocytes involving voltage gated Ca2 + channel , ryanodine receptor and buffers," Alexandria J. Med., vol. 52, no. MARCH, pp. 43–49, 2016.

[20] P. A. Naik and K. R. Pardasani, "Three-Dimensional finite element model to study effect of RyR calcium channel , ER leak and SERCA pump on calcium distribution in oocyte cell," Int. J. Comput. Methods, vol. 15, no. 3, pp. 1–19, 2018.

[21] S. Panday and K. R. Pardasani, "Finite element model to study the mechanics of calcium regulation in oocyte," J. Mech. Med. Biol., vol. 14, no. 2, pp. 1–16, 2014.

[22] K. Pathak and N. Adlakha, "Finite element model to study two dimensional unsteady state calcium distribution in cardiac myocytes," Alexandria J. Med., vol. 52, no. 3, pp. 261–268, 2016.

[23] N. Singh and N. Adlakha, "A mathematical model for interdependent calcium and inositol 1,4,5-trisphosphate in cardiac myocyte," Netw. Model. Anal. Heal. Informatics Bioinforma., vol. 8, no. 1, pp. 1–15, Jul. 2019.

[24] W. Chen, G. Aistrup, J. Andrew Wasserstrom, and Y. Shiferaw, "A mathematical model of spontaneous calcium release in cardiac myocytes," Am. J. Physiol. - Hear. Circ. Physiol., vol. 300, no. 5, May 2011.

[25] H. Kumar, P. A. Naik, and K. R. Pardasani, "Finite element model to study calcium distribution in T Lymphocyte involving Buffers and ryanodine receptors," Proc. Natl. Acad. Sci. India Sect. A Phys. Sci., vol. 88, no. 4, pp. 585–590, 2018.

[26] P. A. Naik, "Modeling the mechanics of calcium regulation in T lymphocyte: A finite element method approach," Int. J. Biomath., vol. 13, no. 5, Jun. 2020.

[27] P. A. Naik and J. Zu, "Modeling and simulation of spatial-temporal calcium distribution in T lymphocyte cell by using a reaction-diffusion equation," J. Bioinform. Comput. Biol., vol. 18, no. 2, May 2020.

[28] Y. Jagtap and N. Adlakha, "Numerical study of one-dimensional buffered advection–diffusion of calcium and IP 3 in a hepatocyte cell," Netw. Model. Anal. Heal. Informatics Bioinforma., vol. 8, no. 1, pp. 1–9, Nov. 2019.

[29] N. Manhas and N. Anbazhagan, "A mathematical model of intricate calcium dynamics and modulation of calcium signalling by mitochondria in pancreatic acinar cells," Chaos, Solitons and Fractals, vol. 145, p. 110741, Apr. 2021.

[30] D. D. Dave and B. K. Jha, "Analytically depicting the calcium diffusion for Alzheimer's affected cell," Int. J. Biomath., vol. 11, no. 07, p. 1850088, 2018.

[31] D. D. Dave and B. K. Jha, "Delineation of calcium diffusion in alzheimeric brain," J. Mech. Med. Biol., vol. 18, no. 2, pp. 1–15, 2018.

[32] M. Bertsch, B. Franchi, N. Marcello, M. C. Tesi, and A. Tosin, "Alzheimer's disease: A mathematical model for onset and progression," Math. Med. Biol., vol. 34, no. 2, pp. 193–214, Jun. 2017.

[33] B. K. Jha, H. Joshi, and D. D. Dave, "Portraying the effect of Calcium-Binding proteins on cytosolic calcium concentration distribution fractionally in nerve cells," Interdiscip. Sci., vol. 10, no. 4, pp. 674–685, 2018.

[34] H. Joshi and B. K. Jha, "Modeling the spatiotemporal intracellular calcium dynamics in nerve cell with strong memory effects," Int. J. Nonlinear Sci. Numer. Simul., vol. 0, no. 0, p. 000010151520200254, Aug. 2021. https://doi.org/10.1515/ijnsns-2020-0254

[35] H. Joshi and B. K. Jha, "On a reaction–diffusion model for calcium dynamics in neurons with Mittag–Leffler memory," Eur. Phys. J. Plus, vol. 136, no. 6, pp. 1–15, Jun. 2021.

[36] H. Joshi and B. K. Jha, "Fractionally delineate the neuroprotective function of calbindin-28k in Parkinson's disease," Int. J. Biomath., vol. 11, no. 08, p. 1850103, 2018.

[37] H. Joshi and B. K. Jha, "Generalized diffusion characteristics of calcium model with concentration and memory of cells: A spatiotemporal approach," Iran. J. Sci. Technol. Trans. A Sci., pp. 1–14, 2021. https://doi.org/10.1007/s40995-021-01247-5

[38] S. Bakshi, V. Chelliah, C. Chen, and P. H. van der Graaf, "Mathematical biology models of Parkinson's disease," CPT: Pharmacometrics and Systems Pharmacology, vol. 8, no. 2. John Wiley & Sons, Ltd, pp. 77–86, 01-Feb-2019.

[39] H. Joshi and B. K. Jha, "Fractional-order mathematical model for calcium distribution in nerve cells," Comput. Appl. Math., vol. 39, no. 2, pp. 1–22, May 2020.

[40] U. Kharde, K. Takale, and S. Gaikwad, "Numerical solution of time fractional drug concentration equation in central nervous system," J. Math. Comput. Sci., vol. 11, no. 6, pp. 7317–7336, 2021.

[41] Q. Lu, "Dynamics and coupling of fractional-order models of the motor cortex and central pattern generators," J. Neural Eng., vol. 17, no. 3, p. 036021, Jun. 2020.

[42] M. Yavuz and A. Yokus, "Analytical and numerical approaches to nerve impulse model of fractional-order," Numer. Methods Partial Differ. Equ., vol. 36, no. 6, pp. 1348–1368, Nov. 2020.

[43] I. Podlubny, Fractional differential equations: an introduction to fractional derivatives, fractional differential equations, to methods of their solution and some of their, 1st Editio. Academic Press, Elsevier, 1998.

[44] K. Diethelm, The analysis of fractional differential equations: an application-oriented exposition using differential operators of Caputo type. Springer-Verlag Berlin Heidelberg, 2010.

[45] S. G. Samko, A. A. Kilbas, and O. I. Marichev, "Fractional integrals and derivatives: Theory and applications," Gordon and Breach Science Publishers. pp. 1–973, 1993.

[46] K. S. Miller and B. Ross, An introduction to the fractional calculus and fractional differential equations. 1993.

[47] R. L. Magin, Fractional Calculus in Bioengineering. Begell House, 2006.

[48] J. Crank, The Mathematics of Diffusion, Second. Oxford University Press, Ely House, London W.I, 1975.

[49] G. D. Smith, L. Dai, R. M. Miura, and A. Sherman, "Asymptotic analysis of buffered calcium diffusion near a point source," SIAM J. Appl. Math., vol. 61, no. 5, pp. 1816–1838, 2001.

[50] G. D. Smith, "Analytical steady-state solution to the rapid buffering approximation near an open Ca2+ channel.," Biophys. J., vol. 71, no. 6, pp. 3064–72, 1996.

12

Approximate Analytic Solution for Tumour Growth and Human Head Heat Distribution Singular Boundary Value Model by High-Resolution Order-Preserving Fuzzy Transform

Navnit Jha and Kritika

Faculty of Mathematics and Computer Science
South Asian University, Delhi, India
E-mail: navnitjha@sau.ac.in

Abstract

The mathematical models describing tumor growth, distribution of heat sources in the human head, and spherical cell oxygen diffusion with Michaelis–Menten uptake kinetics are treated by order-preserving fuzzy transform approximations. The fuzzy transform using a triangular base is applied to obtain approximate fuzzy components. A sequence of functional approximations and modifications in derivative values results in a set of equations that can be easily programmed to receive the approximate numerical solution. Further, the cubic spline interpolation yields an approximate solution in the analytic form. The present formulation offers an approximate analytic solution possessing an order-preserving character. The robustness and accuracy of fuzzy component schemes are measured for singular and non-singular second-order differential equations whose exact solutions are known. The numerical simulations determine the computational order and integrate absolute error to corroborate the applicability of a high-resolution scheme.

Keywords: Fuzzy transform, triangular base, singular differential equations, tumor growth model, high-resolution, computational convergence rate.

12.1 Introduction

The second-order nonlinear differential equations play an important role in the simulation of physiology modeling. Many nonlinear models do not possess exact solutions, and thus, it is essential to treat them by approximation technique. In recent years, the hybrid scheme has been more suitable among the available independent solution methods: finite-difference, wavelets, homotopy analysis, spectral, and spline collocation. One such hybrid technique is fuzzy transform combined with compact discretization, which offers approximate analytic and numerical solutions, especially when the model exhibits singularity inside the physical domain with inherent computational complexity. The fuzzy transform is an elegant tool to treat differential equations by employing a suitable base function on a uniform fuzzy partition of an integral domain. The essence of the fuzzy transform lies in the fact that it captures the solution values in a small subdomain instead of a particular nodal point. Apart from several inherent beauty, the fuzzy transform approximations lack the order-preserving character of solution values. In other words, the accuracy of solutions with the desired precision takes an extensive computing time. Thus, it is cumbersome to implement the fuzzy transform technique to higher dimension differential equations. We can overcome such difficulty by combining the fuzzy transform with finite-difference discretization to achieve more precise solution values in a short computer processing time. Our approach will be based on the three-point approximations of fuzzy components with the help of fuzzy transformation on a uniform fuzzy partition. The difference between approximated and exact fuzzy components helps determine the linear combinations with suitable accuracies. A similar procedure will be followed to solution derivative, and functional approximations along with approximated fuzzy components yield a hybrid scheme with fourth-order accuracy. The resulting scheme yields in a tri-diagonal Jacobian matrix and is easily computed in a memory-efficient manner.

The fuzzy transform in the context of differential equations was described in a lucid manner by Perfilieva [1] and Khastan et al. [2]. Later, the procedure is extended by putting modifications in terms of orthogonal basis, least square, higher degree fuzzy transform [3, 4], and the higher derivatives by fuzzy transform are elaborated in [5]. Fuzzy transform combined with shooting technique for nonlinear boundary value problems is obtained in [6]. Chen and Shen [7] implemented fuzzy transform for second-order differential equations and computed approximate solutions with graphical illustrations.

An alternative to the approximation technique by the neural network, radial basis function, and fuzzy transform is described by Štěpnička and Polakovič [8]. In the present work, the fuzzy transform of a continuous function is evaluated at each element of an exponential basis, which offers an advantage over a polynomial basis since the exponential basis is treated as a generalization over a polynomial basis and both resemble the same solution space in the limiting case of frequency parameter. Also, the oscillations in the numerical solution can often be controlled by tuning the frequency parameter depending upon their locations. A detailed discussion on an exponential basis in the context of the higher-dimensions convection-diffusion-reaction model is presented by Jha and Singh [9]. In the present work, we will consider the nonlinear singular equation

$$\frac{d^2 x(t)}{dt^2} = \phi\left(t, x\,(t),\, \frac{dx(t)}{dt}\right), \quad 0 < t < 1, \tag{12.1}$$

with the associated boundary values $x\,(0) = x_0$, $x\,(1) = x_{N+1}$. The solution's existence and uniqueness are ensured under the assumptions: $\phi\,(t, x, y)$ is continuous inside the domain $\{(t, x, y)\} \mid 0 \le t \le 1,\ -\infty < x, y < \infty\}$, $y = x'\,(t) = dx(t)/dt$, and for $(t, x^*, y),\, (t, x, y^*) \in \{(t, x, y) \mid 0 \le t \le 1,\ -\infty < x, y < \infty\}$, it satisfy Lipschitz conditions $|\phi\,(t, x, y) - \phi\,(t, x^*, y)| \le \omega_1\,|x - x^*|$, $|\phi\,(t, x, y) - \phi\,(t, x, y^*)| \le \omega_2\,|y - y^*|$, along with $\partial\phi/\partial x > 0$, $|\partial\phi/\partial y| \le \omega_3$, where $\omega_1, \omega_2, \omega_3$ are positive constants. The numerical solution of (12.1) employing a spline scheme on a non-uniform nodal spacing was obtained earlier by Jha [10] in the context of physiological models. Asaithambi and Goodman [11] obtained pointwise error bounds for the physiological singular diffusion equation. In the present high-resolution fuzzy component formulation, the outcome of the order-preserving stable numerical and piecewise analytic solution yields an added advantage. The main motivation in the present work is obtaining the approximate analytic solution to the models whose analytic solution is not known. Tumor growth, oxygen diffusion in a spherical cell, and human heat conduction models are singular boundary value problems whose theoretical solutions are unknown. The determination of high-resolution approximate analytic solutions will be of practical importance.

The plan of the chapter is arranged in the following sequence. The next section describes the fuzzy components using the triangular base and corresponding exponential basis fuzzy component approximations for the solution, and solution derivatives are obtained in Section 3. In Section 4, it is extended to the order-preserving fuzzy component scheme. The implantation of a high-resolution fuzzy component scheme is considered for tumor growth and oxygen diffusion in the spherical cell model in section 5, and discretization of the heat distribution model in the human head is obtained in section 6. The detailed numerical simulations and performance evaluation is described in section 7. The chapter is finally concluded with remarks and future scope.

12.2 Preliminaries of Fuzzy Transform

Let N be a given positive integer and $\{t_k=a+kh:k=0,\cdots,\overline{N+1}\}$, be the uniformly spaced nodal points inside the domain $\Omega=[a, b]$, $h=(b-a)/(N+1)$. The fuzzy transform (F-transform) of a real-valued continuous function $x(t)$ on the domain Ω is defined as $F\{x(t)\}$ $=[X_0, X_1,\cdots,X_{N+1}]$, where

$$X_k = \frac{\int_{t_{k-1}}^{t_{k+1}} x(t) A_k(t) \ dt}{\int_{t_{k-1}}^{t_{k+1}} A_k(t) \ dt}, \quad k=1,\cdots,N, \quad X_0 = \frac{\int_{t_0}^{t_1} x(t) A_0(t) \ dt}{\int_{t_0}^{t_1} A_0(t) \ dt},$$

$$X_{N+1} = \frac{\int_{t_N}^{t_{N+1}} x(t) A_{N+1}(t) \ dt}{\int_{t_N}^{t_{N+1}} A_{N+1}(t) \ dt}, \tag{12.2}$$

and

$$A_k(t) = \begin{cases} \frac{t-t_{k-1}}{h}, & t_{k-1} \le t \le t_k \\ \frac{t_{k+1}-t}{h}, & t_k \le t \le t_{k+1} \end{cases} \quad ;k=1,\cdots,N,$$

$$A_0(x) = \begin{cases} \frac{t_1-t}{h}, & t \in [t_0, t_1], \\ 0, & otherwise, \end{cases} \quad A_{N+1}(x) = \begin{cases} \frac{t-t_N}{h}, & t \in [t_N, t_{N+1}], \\ 0, & otherwise. \end{cases}$$
$$\tag{12.3}$$

are the triangular base function that forms a fuzzy partition of the domain Ω. A detailed description of fuzzy partition and base functions in the context of differential equations is given in [1, 2, 4]. The inverse F-transform $\hat{X}(t) = \sum_{k=0}^{N+1} X_k A_k(t)$, yields the approximate solution on the domain Ω. It is observed that, for a continuously differentiable function $x(t)$, the interior fuzzy components are second-order accurate while the boundary fuzzy components are first-order accurate with the corresponding exact solution values. This can be easily seen from the definition (12.2) in the following manner

$$X_0 = \frac{2}{h}\left(\frac{h}{2}x(t_0)+\frac{h^2}{6}x'(t_0)+O(h^3)\right) = x(t_0)+O(h), \tag{12.4}$$

$$X_{N+1} = \frac{2}{h}\left(\frac{h}{2}x\left(t_{N+1}\right) - \frac{h^2}{6}x'\left(t_{N+1}\right) + O(h^3)\right) = x(t_{N+1}) + O(h), \quad (12.5)$$

and

$$X_k = \left[\int_{t_{k-1}}^{t_k} x(t)\, A_k(t)\ dt + \int_{t_k}^{t_{k+1}} x(t)\, A_k(t)\ dt\right] \bigg/$$

$$\left[\int_{t_{k-1}}^{t_k} A_k(t)\ dt + \int_{t_k}^{t_{k+1}} A_k(t)\ dt\right]$$

$$= x(t_k) + \frac{h^2}{12}x'(t_k) + O(h^4) = x(t_k) + O(h^2), \quad k = 1(1)N. \quad (12.6)$$

Further, the inverse F-transform

$$\hat{X}(t) = \sum_{k=0}^{N+1} X_k A_k(t) = X_0 A_0(t) + \sum_{k=1}^{N} X_k A_k(t) + X_{N+1} A_{N+1}(t).$$

$$(12.7)$$

yields the following point-wise approximations

$$\begin{bmatrix} \hat{X}(t_0) \\ \hat{X}(t_1) \\ \vdots \\ \hat{X}(t_N) \\ \hat{X}(t_{N+1}) \end{bmatrix} = \begin{bmatrix} X_0 A_0(t_0) \\ X_1 A_1(t_1) \\ \vdots \\ X_N A_N(t_N) \\ X_{N+1} A_{N+1}(t_{N+1}) \end{bmatrix} \approx \begin{bmatrix} x(t_0) \\ x(t_1) \\ \vdots \\ x(t_N) \\ x(t_{N+1}) \end{bmatrix} + \begin{bmatrix} O(h) \\ O(h^2) \\ \vdots \\ O(h^2) \\ O(h) \end{bmatrix}.$$

$$(12.8)$$

With the specified boundary data, the $O(h)$-accuracy to terminal fuzzy components $\hat{X}(t_0)$ and $\hat{X}(t_{N+1})$ are irrelevant, but the interior fuzzy approximations $\hat{X}(t_1),\cdots,\hat{X}(t_N)$ offer only $O(h^2)$-accurate solution. The second-order fuzzy transform technique consumes enormous computing time when applied to multi-dimensional models or problems possessing singularity. Thus, we aim to improve the order of accuracy of the fuzzy transform scheme (12.8) utilizing the approximated fuzzy components, resulting in a more accurate solution with a short computing time.

12.3 Exponential Basis Approximated Fuzzy Components

In this section, we shall approximate the fuzzy component of $x(t)$ and $x'(t)$ at three neighbouring nodes $t_{k+\varsigma}$, $\varsigma = 0, \pm1$, and the coefficients of the linear combination are evaluated by employing the exponential basis $\mathcal{B} = \{1, e^{\nu t}, e^{-\nu t}\}$, ν being frequency parameter. Consider the following three-point linear combinations

$$
\begin{bmatrix} \overline{X}_k \\ \overline{X}_{k+1} \\ \overline{X}_{k-1} \end{bmatrix} = \begin{bmatrix} p_{k-1}^{(0)} & p_k^{(0)} & p_{k+1}^{(0)} \\ p_{k-1}^{(1)} & p_k^{(1)} & p_{k+1}^{(1)} \\ p_{k-1}^{(2)} & p_k^{(2)} & p_{k+1}^{(2)} \end{bmatrix} \begin{bmatrix} x_{k-1} \\ x_k \\ x_{k+1} \end{bmatrix},
\tag{12.9}
$$

Upon evaluating the exact fuzzy transform X_k defined in (12.2) and approximate fuzzy component $\overline{X}_k$ on each element of the basis $\mathcal{B}$, we get three equations involving three unknowns $p_k^{(0)}$, $p_{k\pm1}^{(0)}$, and set of equations yield the solution

$$
\begin{bmatrix} p_k^{(0)} \\ p_{k+1}^{(0)} \\ p_{k-1}^{(0)} \end{bmatrix} = \frac{1}{\nu^2 h^2 (e^{\nu h}-1)^2} \begin{bmatrix} \nu^2 h^2 \left(e^{2\nu h}+1\right) - 2(e^{\nu h}-1)^2 \\ (e^{\nu h}-1)^2 - \nu^2 h^2 e^{\nu h} \\ (e^{\nu h}-1)^2 - \nu^2 h^2 e^{\nu h} \end{bmatrix}.
\tag{12.10}
$$

With these values, it is easy to verify that $\overline{X}_k - X_k = O(h^4)$. By using the properties of fuzzy partition (for the details, one can refer Perfilieva [4]), we obtain

$$
\begin{aligned}
X_{k+1} &= \frac{\int_{t_k}^{t_{k+2}} x(t) A_{k+1}(t)\, dt}{\int_{t_k}^{t_{k+2}} A_{k+1}(t)\, dt} = \frac{\int_{t_k}^{t_{k+2}} x(t) A_k(t-h)\, dt}{\int_{t_k}^{t_{k+2}} A_k(t-h)\, dt} \\
&= \frac{\int_{t_{k-1}}^{t_{k+1}} x(t+h) A_k(t)\, dt}{\int_{t_{k-1}}^{t_{k+1}} A_k(t)\, dt},
\end{aligned}
\tag{12.11}
$$

$$
\begin{aligned}
X_{k-1} &= \frac{\int_{t_{k-2}}^{t_k} x(t) A_{k-1}(t)\, dt}{\int_{t_{k-2}}^{t_k} A_{k-1}(t)\, dt} = \frac{\int_{t_{k-2}}^{t_k} x(t) A_k(t+h)\, dt}{\int_{t_{k-2}}^{t_k} A_k(t+h)\, dt} \\
&= \frac{\int_{t_{k-1}}^{t_{k+1}} x(t-h) A_k(t)\, dt}{\int_{t_{k-1}}^{t_{k+1}} A_k(t)\, dt}.
\end{aligned}
\tag{12.12}
$$

Now, using the similar derivations discussed above, the values of other parameters are given by

$$
\begin{bmatrix} p_k^{(1)} \\ p_{k+1}^{(1)} \\ p_{k-1}^{(1)} \end{bmatrix} = \begin{bmatrix} p_k^{(2)} \\ p_{k-1}^{(2)} \\ p_{k+1}^{(2)} \end{bmatrix} = \frac{1}{\nu^2 h^2 (e^{\nu h}-1)^2}
$$

$$
\begin{bmatrix} (\nu^2 h^2 + 2)\left(e^{2\nu h}+1\right) - e^{\nu h}\left(2 + e^{2\nu h}\right) - e^{-\nu h} \\ \left(e^{4\nu h}+1\right) e^{-\nu h} - (\nu^2 h^2 - 4) e^{\nu h} - 3\left(e^{2\nu h}+1\right) \\ (e^{\nu h}-1)^2 - \nu^2 h^2 e^{\nu h} \end{bmatrix}.
$$

and it yields $\overline{X}_{k\pm1}-X_{k\pm1}=O(h^3)$. As a result, we get

$$
\begin{bmatrix} \overline{X}_k \\ \overline{X}_{k+1} \\ \overline{X}_{k-1} \end{bmatrix} = \begin{bmatrix} x_k \\ x_{k+1} \\ x_{k-1} \end{bmatrix} + \frac{h^3}{12} \begin{bmatrix} \frac{h}{20}(\nu^2 x_k'' - x_k^{iv}) \\ x_{k+1}''' - \nu^2 x_{k+1}' \\ \nu^2 x_{k-1}' - x_{k-1}''' \end{bmatrix} + \begin{bmatrix} O(h^5) \\ O(h^4) \\ O(h^4) \end{bmatrix}. \qquad (12.13)
$$

We shall now find approximated fuzzy component for the first-order derivative at the three consecutive nodes $t_{k+\zeta}$, $\zeta= 0, \pm1$. The fuzzy component for $x'(t) = dx(t)/dt$ is represented by

$$
\mathcal{F}\left[x\left(t\right)\right] = \left[X_1', X_2', \cdots, X_N'\right], \quad X_k' = \frac{\int_{t_{k-1}}^{t_{k+1}} x'(t) A_k(t)\, dt}{\int_{t_{k-1}}^{t_{k+1}} A_k(t)\, dt}, \quad k= 1,\cdots,N.
$$

$$(12.14)$$

Let us define the approximated fuzzy component for $x'(t)$ as

$$
\begin{bmatrix} \overline{X}_k' \\ \overline{X}_{k+1}' \\ \overline{X}_{k-1}' \end{bmatrix} = \begin{bmatrix} q_{k-1}^{(0)} & q_k^{(0)} & q_{k+1}^{(0)} \\ q_{k-1}^{(1)} & q_k^{(1)} & q_{k+1}^{(1)} \\ q_{k-1}^{(2)} & q_k^{(2)} & q_{k+1}^{(2)} \end{bmatrix} \begin{bmatrix} x_{k-1} \\ x_k \\ x_{k+1} \end{bmatrix}. \qquad (12.15)
$$

The evaluation of the difference between approximate and exact fuzzy components of derivative at three neighbouring nodes $\begin{bmatrix} \overline{X}_k' & \overline{X}_{k+1}' & \overline{X}_{k-1}' \end{bmatrix}^T - \begin{bmatrix} X_k' & X_{k+1}' & X_{k-1}' \end{bmatrix}^T$, on each element of the basis $\mathcal{B}=\{1, e^{\nu t}, e^{-\nu t}\}$ yields the following values of parameters

$$
\begin{bmatrix} q_k^{(0)} \\ q_{k+1}^{(0)} \\ q_{k-1}^{(0)} \end{bmatrix} = \frac{1}{\omega} \begin{bmatrix} 0 \\ e^{\nu h}-1 \\ 1-e^{\nu h} \end{bmatrix}, \quad \begin{bmatrix} q_k^{(1)} \\ q_{k+1}^{(1)} \\ q_{k-1}^{(1)} \end{bmatrix} = -\begin{bmatrix} q_k^{(2)} \\ q_{k-1}^{(2)} \\ q_{k+1}^{(2)} \end{bmatrix}
$$

$$
= \frac{1}{\omega} \begin{bmatrix} e^{-\nu h}\left(1-e^{2\nu h}\right)\left(e^{\nu h}+1\right) \\ e^{-\nu h}\left(e^{3\nu h}-1\right) \\ e^{\nu h}-1 \end{bmatrix}, \qquad (12.16)
$$

where $\omega=\nu h^2\left(e^{\nu h}+1\right)$. Now, it is easy to see that the approximated fuzzy components of derivatives

$$
\begin{bmatrix} \overline{X}_k' \\ \overline{X}_{k+1}' \\ \overline{X}_{k-1}' \end{bmatrix} = \begin{bmatrix} x_k' \\ x_{k+1}' \\ x_{k-1}' \end{bmatrix} + \frac{h^2}{12} \begin{bmatrix} 2x_k''' - \nu^2 x_k' \\ -4x_k''' + 5\nu^2 x_k' \\ -4x_k''' + 5\nu^2 x_k' \end{bmatrix} + \begin{bmatrix} O(h^3) \\ O(h^3) \\ O(h^3) \end{bmatrix}. \qquad (12.17)
$$

These approximate fuzzy components are used to improve the order of the scheme so that we can achieve $O(h^4)$-accuracy to the nonlinear equation (12.1).

12.4 Order Preserving Fuzzy Component Scheme

We shall describe an order-preserving three-point discretization for solving the nonlinear equation (12.1). The approximations of functions are carried out on the fuzzy components defined by the relation (12.9) and (12.15), and Taylor's expansion helps us to achieve the local truncation error of sixth-order. Let us define

$$\overline{\phi}_k = \phi(t_k, \overline{X}_k, \overline{X}'_k), \tag{12.18}$$

$$\overline{\phi}_{k+1} = \phi(t_{k+1}, \overline{X}_{k+1}, \overline{X}'_{k+1}), \tag{12.19}$$

$$\overline{\phi}_{k-1} = \phi(t_{k-1}, \overline{X}_{k-1}, \overline{X}'_{k-1}), \tag{12.20}$$

Upon the substitution of (12.13) and (12.17) in (12.18)-(12.20), the series expansion yields

$$\overline{\phi}_k = \phi_k - \frac{h^2}{6}\left(\frac{1}{2}\nu^2 B_k x'_k - \frac{1}{2}A_k x''_k - B_k x'''_k\right) + O(h^4), \tag{12.21}$$

$$\overline{\phi}_{k+1} = \phi_{k+1} + \frac{h^2}{12}\Big[\nu^2(5B_k + (5B'_k + A_k)h)x'_k + (A_k + hA'_k)x''_k$$
$$- 4(B_k + hB'_k)x'''_k\Big] + O(h^4), \tag{12.22}$$

$$\overline{\phi}_{k-1} = \phi_{k-1} + \frac{h^2}{12}\Big[\nu^2(5B_k - (5B'_k + A_k)h)x'_k + (A_k - hA'_k)x''_k$$
$$- 4(B_k - hB'_k)x'''_k\Big] + O(h^4), \tag{12.23}$$

where

$$\phi_k = \phi_k\left(t_k, x\left(t_k\right), x'\left(t_k\right)\right), \phi_{k\pm1} = \phi_{k\pm1}\left(t_{k\pm1}, x\left(t_{k\pm1}\right), x'\left(t_{k\pm1}\right)\right),$$

$$A\left(t\right) = \frac{\partial\phi}{\partial x(t)}, B\left(t\right) = \frac{\partial\phi}{\partial x'(t)}, A_k = A\left(t_k\right),$$

$$A'_k = \frac{\partial A}{\partial t}\bigg|_{t=t_k}, B_k = B\left(t_k\right), B'_k = \frac{\partial B}{\partial t}\bigg|_{t=t_k}.$$

Similar to the classical Cowell's type method (with required modification on an exponential basis), the fuzzy component combined with compact formulation is given by

$$\overline{X}_{k-1} - 2\overline{X}_k + \overline{X}_{k+1} = h^2\left(c_k\,\overline{\phi}_k + c_{k-1}\,\overline{\phi}_{k-1} + c_{k+1}\,\overline{\phi}_{k+1}\right) + T_k^{(1)}, \tag{12.24}$$

where

$$c_{k-1} = c_{k+1} = \frac{1}{12}\left(1 + \frac{1}{12}\nu^2 h^2\right), \; c_k = \frac{5}{6}\left(1 + \frac{1}{12}\nu^2 h^2\right).$$

Upon clubbing (12.21)–(12.23) in (12.24) and carrying out necessary algebra, we get

$$T_k^{(1)} = -\frac{h^4}{12}(A_k x_k'' + B_k x_k''') + O(h^6).$$

The local truncation error $T_k^{(1)} = O(h^4)$, and it is possible to refine the approximation so that the error can be brought down to $O(h^6)$, $0 < h \ll 1$. This can be accomplished by updating the functional approximations $\bar{\phi}_k$ defined at the central node, thus assuming

$$\bar{\bar{X}}_k = \bar{X}_k + h^2 \alpha(\bar{\phi}_{k-1} + \bar{\phi}_{k+1}), \bar{\bar{X}}_k' = \bar{X}_k' - h\alpha(\bar{\phi}_{k-1} - \bar{\phi}_{k+1}), \quad (12.25)$$

$$\bar{\bar{\phi}}_k = \phi(t_k, \bar{\bar{X}}_k, \bar{\bar{X}}_k'). \quad (12.26)$$

Using the series expansion on (12.26), we obtain

$$\bar{\bar{\phi}}_k = \phi_k - \frac{\nu^2 h^2}{12}B_k x_k' + \frac{h^2}{12}(1 + 24\alpha)A_k x_k'' + \frac{h^2}{6}(12\alpha+1)x_k''' + O(h^4). \quad (12.27)$$

Now, implementing the relation (12.24) on the updated functional values, we obtain

$$\bar{X}_{k-1} + \bar{X}_{k+1} - 2\bar{X}_k = h^2\left(c_{k-1}\bar{\bar{\phi}}_{k-1} + c_k\bar{\bar{\phi}}_k + c_{k+1}\bar{\bar{\phi}}_{k+1}\right) + T_k^{(2)}. \quad (12.28)$$

where

$$\alpha = -\frac{1}{20}$$

and

$$T_k^{(2)} = \frac{h^6}{720}\left(2\nu^2 x_k'' - 3x_k'''\right) + O(h^7)$$

.

The $O(h^6)$ accuracy of $T_k^{(2)}$ remains intact whether the frequency parameter ν approaches zero or not. However, in case $\nu \to 0$, the scheme (12.28) is a fourth-order compact formulation on a polynomial basis $\{1, x, x^2\}$. The implementation of an exponential basis high-resolution fuzzy component scheme (12.28) requires modification for the model possessing singular terms. A detailed derivation for the singular model is described in the following sections.

12.5 Tumor Growth and Oxygen Diffusion in a Spherical Cell Model

Tissue stability is a complex biological phenomenon in tumor growth, and a mathematical model is an elegant mechanism to understand the complicated growth of tissue. The primary purpose of mathematical models is to describe the qualitative development of tumor tissue in the early stage under prevailing assumptions. The nonlinear singular differential equation

$$x''(t) + \frac{m}{t}x'(t) = \frac{\eta x(t)}{\mu + x(t)}, \quad \eta, \mu > 0, \ 0 < t < 1, \ m = 0, 1, 2, \quad (12.29)$$

appears in the modeling and simulation of different physiological phenomena such as tumor growth and oxygen diffusion inside a spherical call with Michaelis-Menten uptake [12, 13]. In the case of oxygen, diffusion $x(t)$ defines oxygen tension, $t(=r)$ is the radial distance, η is the diffusion coefficient and μ denotes the maximum reaction rate [14]. The analytic solution to the nonlinear model (12.29) is not known, thus, an approximate analytic or numerical solution is essential to analyze the solution behavior. Moreover, if $m \neq 0$, the model exhibits singularity at $t = 0$ and consequently the difficulties arise in the numerical solution. Here, we shall present a compact operator technique that takes care of singular term and computation does not halt at $t = 0$. The application of scheme (12.28) to the model (12.29) for

$$\phi\left(t, x(t), x'(t)\right) = -\frac{m}{t}x'(t) + \frac{\eta x(t)}{\mu + x(t)}, \quad (12.30)$$

yields the $O(h^4)$-accurate fuzzy component scheme that appears with terms $\frac{1}{t_{k-1}}$. At $k = 1$, $\frac{1}{t_{k-1}} = \frac{1}{t_0}$ and $t_0 = 0$ halts the execution. Such a problem can be resolved by expressing $x_{k\pm1}$ in terms of operators. Let

$$x_{k+1} = \frac{1}{2}(2 + \delta^{(1)} + \delta^2)x_k \text{ and } x_{k-1} = \frac{1}{2}(2 - \delta^{(1)} + \delta^2)x_k. \quad (12.31)$$

Alternatively, the operators

$$\delta^{(2)}x_k = x_{k+1} + x_{k-1} - 2x_k = h^2 x''(t) + O(h^4) \quad (12.32)$$

and

$$\delta^{(1)}x_k = x_{k+1} - x_{k-1} = 2hx'(t) + O(h^3), \quad (12.33)$$

estimates $x''(t)$ and $x'(t)$ with the accuracy of $O(h^2)$. Using Taylor's expansion, one can express

$$\frac{1}{t_{k-1}} = \frac{1}{t_k} + \frac{h}{t_k} + \frac{h^2}{t_k} + \frac{h^3}{t_k} + O(h^4). \quad (12.34)$$

Substituting (12.30), (12.31), and (12.34) in (12.28), one obtains

$$\frac{1}{12}\left[h^2\left(p_k^2-2p_k'+\nu^2\right)+12-\frac{2h\eta\mu}{(\mu+x_k)^2}\right]\delta^{(2)}x_k+\frac{h^2\eta\mu\left(\delta^{(1)}x_k\right)^2}{24(\mu+x_k)^3}$$

$$+\frac{h}{24}\left[h^2\left\{p_k\left(p_k'-\nu^2\right)-p_k''\right\}-12p_k+\frac{2h^2\eta\mu p_k}{(\mu+x_k)^2}\right]\delta^{(1)}x_k$$

$$-\frac{h^2}{12}\frac{\eta x_k\left[(\mu+x_k)^2\left(\nu^2h^2+12\right)-\eta\mu h^2\right]}{(\mu+x_k)^3}$$

$$=O\left(h^4\delta^{(2)}x_k\right)+O\left(h^5\delta^{(1)}x_k\right)+O\left(h^6\right),\tag{12.35}$$

where $p(t)=\frac{1}{t}$ and $\left[p_k,p_k',p_k''\right]=\left[p(t_k),p'(t_k),p''(t_k)\right]$. Since $O\left(\delta^{(2)}x_k\right)$ $=O(h^2)$, and $O\left(\delta^{(1)}x_k\right)=O(h)$, therefore right-hand side of (12.35) contributes an $O(h^6)$ local truncation error, and hence can be easily neglected for the numerical simulations. The scheme (12.35) is free from the term $\frac{1}{t_{k-1}}$ and easily computed near the singular point after removing the right-hand side in (12.35). The piece-wise interpolating polynomial obtained using exponential basis high-resolution fuzzy component scheme (12.35) yields the approximate solutions values inside the domain $\Omega=[0,1]$. The approximate piecewise cubic spline polynomial for the solution at $N=6$ is obtained with $\eta=10$, $\mu=2$, $m=1$, and it is given by

$$\hat{x}(t)=\begin{cases}0.82+0.82x+0.79x^2+0.77x^3, & 0.00\leq t<0.14\\0.77+0.75x+0.72x^2+0.70x^3, & 0.14\leq t<0.29\\0.70+0.68x+0.66x^2+0.65x^3, & 0.29\leq t<0.43\\0.65+0.63x+0.63x^2+0.63x^3, & 0.43\leq t<0.57\\0.63+0.63x+0.65x^2+0.67x^3, & 0.57\leq t<0.71\\0.67+0.69x+0.72x^2+0.77x^3, & 0.71\leq t<0.86\\0.77+0.82x+0.87x^2+0.95x^3, & 0.86\leq t<1.00\end{cases}\tag{12.36}$$

The evaluation of (12.36) is obtained using the frequency parameter value $\nu=0.5$ and boundary data $x(0)=0.82$ and $x(1)=0.95$, see Rashidinia et al. [15]. The graphical illustration on the two data sets comprising $\nu=0.5$, $\eta=10$, $N=16$, $m=2$ for the range of values $0.01\leq\mu\leq10$, in Figure 12.1 and $\nu=0.5$, $\mu=2$, $N=16$, $m=1$ for the range of values $1\leq\eta\leq11$, in Figure 12.2 are presented to illustrate the nature of solution $x(t)$ over the domain $0\leq t\leq1$.

12.6 Heat Distribution in the Human Head

The effect of environmental temperature on the heat distribution of the human head is described earlier by Flesch [16]. It is observed that the generation of heat in

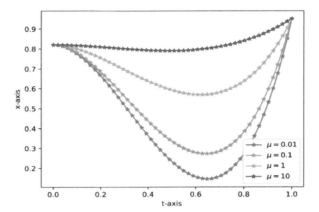

Figure 12.1   Effect of parameter μ on the solution.

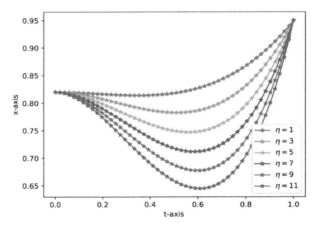

Figure 12.2   Impact of parameter η on the solution.

the surrounding of the human head grows fast compared to the center in response to a diminishing ambient temperature. The mathematical model assuming heat generation is an increasing function of the radial distance from the center, takes the following nonlinear form

$$x''(t) + \frac{2}{t}x'(t) = -de^{-\sigma x(t)}, \quad 0 < t < 1, d > 0, \sigma > 0. \tag{12.37}$$

The presence of a singular term in the model (12.37) makes the computation tedious near the left end $t = 0$. The application of fuzzy component scheme (12.28) for $\phi\left(t, x(t), x'(t)\right) = -2x'(t))/t - de^{-\sigma x(t)}$, yields the discrete equation and employing the similar mechanism discussed in (12.31)-(12.34) results in

$$\left[1+\frac{h^2}{12}\left(p_k^2-p_k'+\nu^2-2d\sigma e^{-\sigma x_k}\right)\right]\delta^{(2)}x_k$$
$$+\frac{h}{24}\left[h^2\left\{p_k\left(2d\sigma e^{-\sigma x_k}-\nu^2+p_k'\right)-p_k''\right\}-12p_k\right]\delta^{(1)}x_k$$
$$+\frac{h^2}{48}d\sigma^2 e^{-\sigma x_k}\left(\delta^{(1)}x_k\right)^2-\frac{h^2}{12}d(h^2\left(e^{-\sigma x_k}d\sigma-\gamma^2\right)e^{-\sigma x_k}-12e^{-\sigma x_k}$$
$$=O\left(h^4\delta^{(2)}x_k\right)+O\left(h^5\delta^{(1)}x_k\right)+O\left(h^6\right), \tag{12.38}$$

The numerical solution of the nonlinear discrete equation (12.38) for $k=1,\cdots,N$ can be obtained with the help of the Newton-Raphson iterative procedure after removing the sixth-order local truncation error term $O\left(h^4\delta^{(2)}x_k\right)+O\left(h^5\delta^{(1)}x_k\right)+O\left(h^6\right)\approx O\left(h^6\right)$. The cubic interpolating polynomial obtained using exponential basis high-resolution fuzzy component yields the point-wise approximate solutions values inside the domain $\Omega=[0,\ 1]$. The approximate piecewise cubic spline polynomial for $N=8$ is obtained with $d=1,\ \sigma=10$, and it is given by

$$\hat{x}(t)=\begin{cases}
0.11+0.17x+0.13x^2+0.14x^3, & 0.00\leq t<0.11\\
0.14+0.14x+0.14x^2+0.14x^3, & 0.11\leq t<0.22\\
0.14+0.14x+0.14x^2+0.14x^3, & 0.22\leq t<0.33\\
0.14+0.14x+0.14x^2+0.14x^3, & 0.33\leq t<0.44\\
0.14+0.14x+0.13x^2+0.13x^3, & 0.44\leq t<0.56\\
0.13+0.13x+0.13x^2+0.13x^3, & 0.56\leq t<0.67\\
0.13+0.12x+0.12x^2+0.12x^3, & 0.67\leq t<0.78\\
0.12+0.12x+0.11x^2+0.11x^3, & 0.78\leq t<0.89\\
0.11+0.11x+0.10x^2+0.10x^3, & 0.89\leq t<1.00
\end{cases} \tag{12.39}$$

The evaluation of (12.39) is obtained using the frequency parameter value $\nu=0.2$ and boundary data $x(0)=0.11$ and $x(1)=0.10$, see Rashidinia [15]. The graphical illustration on the two data sets comprising $\nu=0.1,\sigma=10,N=20$, for the range of values $0.1\leq d\leq 10^3$ in Figure 12.3 and $\nu=0.1,d=1,N=20$, for the range of values $1\leq\sigma\leq 40$, in Figure 12.4 are presented to illustrate the nature of solution $x(t)$ over the domain $0\leq t\leq 1$.

12.7 Numerical Simulations and Performance Evaluation

The numerical techniques are utilized for profound understanding to anticipate the anomalies, which are impractical in the analytical method because they can tackle just a few nonlinear differential equations, but the approximation techniques can precisely apply to the wide range of higher dimensions nonlinear models. The

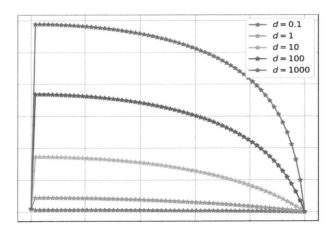

Figure 12.3   Solution behavior for changing d.

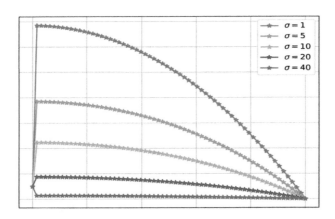

Figure 12.4   Solution behavior for changing σ.

essence of numerical approximations lies in their solution accuracy, computational order, and data storage consumption. We will analyze the numerical solution values $\{x_k\}$ and the approximate analytic solution $\hat{x}(t)$ obtained by employing cubic spline interpolation. The choice $N = 3M+2$, makes it possible to compute $\hat{x}(t)$ for the four consecutive nodal data $t_k, t_{k+1}, t_{k+2}, t_{k+3}, \ k= 0\,(3)\,\overline{N-1}$, by using the Python scipy.interpolate.UnivariateSpline module. Three different metrics will be considered to measure the performance of the fuzzy component scheme. The integrated absolute error quantifying the deviation in approximate analytic solution values and exact solution is defined as

$$E_i^N = \int_a^b |x(t) - \hat{x}(t)| \, dt. \tag{12.40}$$

The maximum-absolute and root-mean-squared errors are defined by

$$\ell_\infty^N = \max_{1 \le k \le N} |x(t_k) - x_k|, \quad \ell_2^N = \left(\frac{1}{N} \sum_{k=1}^N |x(t_k) - x_k|^2 \right)^{1/2}. \tag{12.41}$$

The computational convergence order to these measures are calculated using

$$O_e = log_2 \left(E_i^N / E_i^{2N} \right), \quad O_\infty = log_2 \left(\ell_\infty^N / \ell_\infty^{2N} \right), \quad O_2 = log_2 \left(\ell_2^N / \ell_2^{2N} \right). \tag{12.42}$$

The numerical simulations for linear equations are performed using a tri-diagonal solver, and nonlinear equations are treated with the Newton-Raphson iterative procedure using the initial guess as a zero vector with the error tolerance 10^{-10}. The boundary data are taken from the exact solution unless specified otherwise. Symbolic and numerical computations are implemented separately in Maple 2017 and Python 3.0 on the MacBook Pro 6-Core Intel i7 processor.

Example 12.1 Consider the linear non-singular sewage convection model describing stream concentration at any location

$$x''(t) - \beta x'(t) - \epsilon^2 x(t) = 0, \ 0 < t < l, \tag{12.43}$$

With the preassigned boundary concentration $x(0) = x_0$, $x(l) = x_l$ at the boundary $t = 0$ and $t = l$, the analytic expression satisfying (12.43), (see [17]) is given by

$$x(t) = \left[x_0 e^{l\beta/2} sinh \left(\frac{\theta(l-t)}{2} \right) + x_l sinh \left(\frac{\theta t}{2} \right) \right] \frac{e^{(t-l)\beta/2}}{sin(l\theta/2)},$$

$$\theta = \frac{1}{2} (4\epsilon^2 + \beta^2)^{1/2}. \tag{12.44}$$

The simulations are carried for $l = 5$, $\epsilon = 10$, $\beta = 40$, and boundary concentration $x_0 = 25$, $x_l = 0$. All the three error measures are presented in Table 12.1 at $\nu = 2.4$ for various nodal points. The computational convergence rate comes close to four for a fixed value of the frequency parameter. The plot of each error metric on the log scale is presented in Figure 12.5 within the range $10^{-4} \le \nu \le 6$, showing the effect of the frequency parameter ν on solution accuracy. The minimum error appears a little beyond 10^0 on the log scaled axis. It appears that the scheme that implements an exponential basis is more accurate than the scheme designed on a standard polynomial basis.

Example 12.2 The Bessel differential equation plays a prominent role in x-ray diffraction by molecular crystals, such as the pattern of a helical molecule wrapped

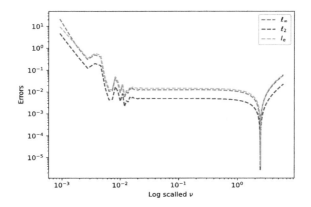

Figure 12.5   Effect of ν on the solution errors .

Table 12.1   Integrated absolute errors and convergence order in example 12.1.

N	E_i^N	O_e	ℓ_2^N	O_2	ℓ_∞^N	O_∞
2	4.86e-01	—	7.76e-02	—	1.90e-01	—
4	9.12e-02	3.0	1.69e-02	2.7	4.06e-02	2.8
8	9.34e-03	3.7	1.66e-03	3.7	4.09e-03	3.7
16	7.35e-04	3.9	1.27e-04	3.9	3.16e-04	3.9

Table 12.2   Integrated absolute errors and convergence order at $\tau = 6$ in example 12.2.

N	E_i^N	O_e	ℓ_2^N	O_2	ℓ_∞^N	O_∞
4	4.47e-01	—	5.03e-02	—	9.03e-02	—
8	2.91e-02	4.4	2.30e-03	5.0	5.81e-03	4.4
16	2.19e-03	4.0	1.79e-04	3.9	4.63e-04	3.9
32	1.55e-04	3.9	1.27e-05	3.9	3.30e-05	3.9
64	1.03e-05	4.0	8.44e-07	4.0	2.23e-06	3.9

around a cylinder of the helix. The Bessel equation also appears in molecular biology for the secondary structures of a helix, particularly proteins and molecular genetics for obtaining the double-helix form of Deoxyribonucleic acid molecular crystals, which has significant outcomes in mutagenesis, medicine, genetics, and evolution of molecules. The Bessel function $x(t) = BesselJ(\tau, t)$ is the exact solution to the equation (12.1) for

$$\phi\left(t, x(t), x'(t)\right) = -\frac{1}{t}x'(t) - \left(1 - \frac{\tau^2}{t^2}\right)x(t). \qquad (12.45)$$

The numerical simulations for $\tau = 6$ on the domain $-5\pi \leq t \leq 5\pi$ is performed for the frequency parameter value $\nu = 0.1$, and solutions errors along with convergence order are reported in Table 12.2. Each error estimate resembles the accuracy close to the theoretical order and precisely computes the solution.

Table 12.3 Maximum absolute and root-mean-squared errors in example 12.3.

N	ℓ_2^N	O_2	ℓ_∞^N	O_∞	ℓ_2^N-error in [19]	ℓ_∞^N-error in [19]
8	3.20e-08	—	4.82e-08	—	7.71e-05	3.89e-05
16	2.89e-09	3.7	4.55e-09	3.6	2.67e-05	8.34e-06
32	1.32e-10	3.7	3.82e-10	3.7	9.38e-06	2.08e-06
64	1.74e-11	3.8	2.99e-11	3.7	3.31e-06	5.20e-07

Table 12.4 Errors and convergence order in example 12.4 for changing ν.

ν	N	ℓ_2^N	O_2	ℓ_∞^N	O_∞
$R_e=10^2$					
0.028	16	5.75e-04	—	4.06e-03	—
0.094	32	5.52e-05	3.5	3.40e-04	3.7
0.492	64	2.89e-06	4.3	1.98e-05	4.2
$R_e=10^3$					
0.007	16	7.16e-02	—	4.35e-01	—
0.015	32	6.33e-03	3.6	5.84e-02	3.0
0.051	64	5.72e-04	3.5	7.54e-03	3.0

Example 12.3 The mathematical model of thermal explosion in a cylindrical vessel is described by singular Lane-Emden equation

$$\phi\left(t, x\left(t\right), x^{'}\left(t\right)\right) = -\frac{1}{t}x^{'}\left(t\right) - e^{-x(t)}, \tag{12.46}$$

having the exact solution $x\left(t\right) = 2\,log\left[(4 - 2\sqrt{2})/\left\{\left(3 - 2\sqrt{2}\right)t^2 + 1\right\}\right]$. The application of fuzzy component scheme (12.28) and employing singularity removing technique discussed in (12.31)-(12.34), on the domain $0 \leq t \leq 1$ for determining solution accuracies is presented in Table 12.3 for $\nu = 0.01$ and various nodal arrangements. In each error metric, the convergence rate is close to four, showing the resemblance of $O(h^4)$-convergence. The proposed scheme shows superiority in terms of ℓ_∞^N and ℓ_2^N-errors with the existing results obtained in Singh et al. [19].

Example 12.4 Burgers equation appears in the biological diffusing system, especially irradiation of tumour tissue [18]. The stationary form of Burgers equation is represented with

$$\phi\left(t, x\left(t\right), x^{'}\left(t\right)\right) = R_e x'(t)\left(x\left(t\right) - 1\right), \tag{12.47}$$

Table 12.5 Integrated absolute errors and convergence order in example 12.5.

N	E_i^N	O_e	ℓ_2^N	O_2	ℓ_∞^N	O_∞
4	4.55e-06	—	6.04e-06	—	8.65e-06	—
8	3.98e-07	3.9	5.38e-07	3.9	8.04e-07	3.8
16	3.04e-08	3.9	4.12e-08	3.9	6.24e-08	3.9
32	1.47e-09	4.5	2.43e-09	4.2	3.93e-09	4.1

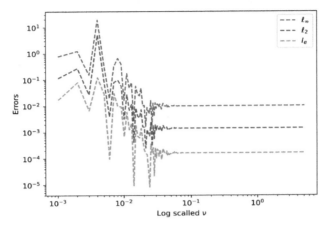

Figure 12.6   Graph of frequency parameter versus errors.

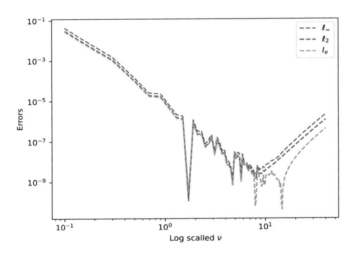

Figure 12.7   Impact of frequency parameter on errors.

and possesses the theoretical solution $x(t) = 1 - tanh\,(R_e t/2)$. The term R_e is known as the Reynolds number. The solution approximations to (12.47) usually deteriorate with a large value of Reynolds number. The frequency parameter in the exponential basis helps in tuning the fourth-order accuracy especially with a high value of R_e. The computational experiment on the domain $0 \le t \le 1$ is performed in Table 12.4 with $R_e = 10^2$ and $R_e = 10^3$ for different values of ν and N. The effect of frequency parameter ν on a log scale is presented in Figure 12.6 for $N = 16$,

$R_e=10^2$ over the range $10^{-3}\leq\nu\leq5$. The deep in the graph is observed between 10^{-2} and 10^{-1}. Such an occurrence beyond 10^0 signifies the importance of taking the non-zero (in the limiting case) frequency parameter value.

Example 12.5 The nonlinear non-singular equation $\phi\left(t, x\left(t\right), x'\left(t\right)\right) = -1 - \left(x'\left(t\right)\right)^2$, is solved over the domain $0\leq t\leq1$ and compared the exact solution $x\left(t\right)=log\left(\cos\left(t-0.5\right)/\cos\left(0.5\right)\right)$ with the numerical solution for the frequency parameter $\nu=8.7$. Results associated with errors are reported in Table 12.5 for different arrangements of nodal points. The solution accuracy is sharply affected by the variation in frequency parameter ν as illustrated in Figure 12.7 over the range $10^{-2}\leq\nu\leq10$ with $N=32$. The downward peak is away from a typically low value of frequency parameter.

12.8 Conclusion and Remarks

The fuzzy component approximation is presented based on three-point compact discretization for the nonlinear singular second-order differential equations. The $O(h^4)$-accuracy of the numerical solutions is analyzed for several physiological models such as tumor growth, oxygen diffusion in the spherical cell, and heat distribution model in humans. The superiority of the present finding is the implementation of a non-polynomial (exponential) basis to determine the fuzzy components, which later helps determine an order-preserving solution scheme. The frequency parameter of an exponential basis can easily be tuned according to the level of oscillations. It is possible to extend such formulations in parabolic partial differential equations that play an important role in modeling time-dependent physiological models.

References

[1] I. Perfilieva, Fuzzy transforms, Transactions on rough sets II. Springer, Berlin, Heidelberg, (2004) 63-81.
[2] A. Khastan, Z. Alijani, I. Perfilieva, Fuzzy transform to approximate solution of two-point boundary value problems, Math. Meth. Appl. Sci. 40 (2017) 6147-6154.
[3] I. Perfilieva, M. Daňková, B. Bede, Towards a higher degree F-transform, Fuzzy Sets Syst. 180 (2011) 3–19.
[4] I. Perfilieva, Fuzzy transforms: Theory and applications, Fuzzy Sets Syst. 157 (2006) 993–1023.
[5] I. Perfilieva, V. Kreinovich, Fuzzy transforms of higher order approximate derivatives: A theorem, Fuzzy Sets Syst. 180 (2011) 55–68.
[6] I. Perfilieva, P. Ševuliáková, R, Valášek, F-transform-based shooting method for nonlinear boundary value problems, Soft Comput. 21 (2017) 3493–3502.

[7] W. Chen, Y. Shen, Approximate solution for a class of second-order ordinary differential equations by the fuzzy transform, J. Intell. Fuzzy Syst. 27 (2014) 73-82.

[8] M. Štěpnička, O. Polakovič, A neural network approach to the fuzzy transform, Fuzzy Sets Syst. 160 (2009) 1037–1047.

[9] N. Jha, B. Singh, Exponential basis and exponential expanding grids third (fourth)-order compact schemes for nonlinear three-dimensional convection-diffusion-reaction equation, Adv. Differ. Equ. 1 (2019) 1-27.

[10] N. Jha, B. K. Mishra, A class of spline schemes for non-linear weakly regular singular boundary value problems using non-uniform mesh: applications to physiological models, Int. J. Math. Model. Simul. Appl. 5 (2012) 180-189.

[11] N.S. Asaithambi, J.B. Goodman, Point wise bounds for a class of singular diffusion problems in physiology, Appl. Math. Comput. 30 (1989) 215–222.

[12] M. Mohsenyzadeh, K. Maleknejad, R. Ezzati, A numerical approach for the solution of a class of singular boundary value problems arising in physiology. Adv Differ. Equ. 231 (2015) 1-10.

[13] H.S. Lin, Oxygen diffusion in a spherical cell with nonlinear oxygen uptake kinetics, J. Theor. Biol. 60(1976) 449–457.

[14] P. Hiltmann and P. Lory On oxygen diffusion in a spherical cell with michaelis-menten oxygen uptake kinetics, Bull. Math. Biol. 45 (1983) 661-664.

[15] J. Rashidinia, R. Mohammadi, R. Jalilian, The numerical solution of non-linear singular boundary value problems arising in physiology, Appl. Math. Comput. 185 (2007) 360–367

[16] U. Flesch, The distribution of heat sources in the human head: a theoretical consideration, J. Theor. Biol. 54 (1975) 285-287

[17] R. P. Agarwal, S. Hodis, D. O'Regan, 500 Examples and Problems of Applied Differential Equations, Springer Nature Switzerland AG (2019).

[18] T. Patel, M.N. Mehta, V.H. Pradhan, The numerical solution of Burger's equation arising into the irradiation of tumour tissue in biological diffusing system by homotopy analysis method, Asian J. Appl. Sci. 5 (2012) 60-66.

[19] R. Singh, H. Garg, V. Guleria, Haar wavelet collocation method for Lane–Emden equations with Dirichlet, Neumann and Neumann–Robin boundary conditions, J. Comput. Appl. Math. 346 (2019) 150–161

13

Analysis of One-Dimensional Groundwater Recharge by Spreading using Hybrid Differential Transform and Finite Difference Method

Aruna Sharma[1*] and Amit K Parikh[2]

[1]Research Scholar, Department of Mathematics, Mehsana Urban Institute of Sciences,Ganpat University, Kherva-384012, Mehsana, Gujarat, India
[2]Principal, Mehsana Urban Institute of Sciences, Ganpat University,Kherva-384012, Mehsana, Gujarat, India
E-mail: sharmaaruna961@gmail.com
*Corresponding Author

Abstract

This study aims to analyze the problem of groundwater recharge in a downward direction. The groundwater is recharged by water spreading vertically downward, resulting in an increase in soil moisture content. The Hybrid Differential Transform and the Finite Difference Method are used to solve the nonlinear partial differential equation that arises as a result of this phenomenon. This method is a combination of the Differential Transform method and Finite Difference Method. The flexibility of the Differential Transform Method is combined with the efficiency of the Finite Difference Method. The numerical solution is obtained using a simple iterative approach, which cuts down on computation time. The numerical solution of the governing equation and the graphical representation has been obtained using MATLAB.

Keywords: Groundwater recharge, Spreading, Burger's equation, Hybrid differential and finite difference method.

13.1 Introduction

The hydrological process of groundwater recharge occurs when water travels downwards from the surface to the groundwater. The tiny open gaps between rock and sand, soil and gravel are where groundwater is stored. It is divided into two zones. The unsaturated zone is made up of rock and sediment whose pore spaces contain large air and a small amount of water and are thus not saturated. Typically, the unsaturated zone begins at the surface and extends downwards to the saturated zone. In the saturated zone, water fills all pores and rock fractures. The water table is at the top of the saturated zone as shown in Figure 13.1.

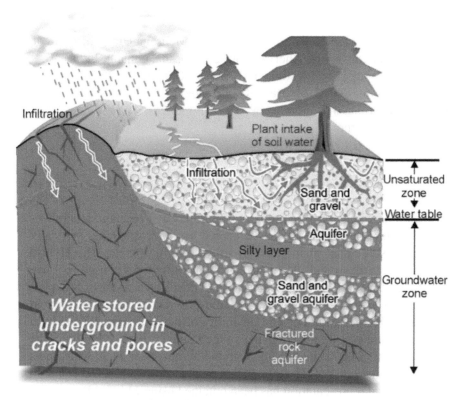

Figure 13.1   Ground water recharge

Source: https://www.gov.nl.ca/ecc/waterres/cycle/groundwater

Soil absorbs and retains water like a sponge. The amount of water in the soil is measured by its moisture content. Infiltrated water flows through the unsaturated region of groundwater and joins the water table. This process is known as groundwater recharge.

Moisture content plays a significant role in a wide range of scientific and technical fields. Because there is no moisture in dry soil, the value is 0; however, when the medium is fully saturated with water, the value is 1. As a result, moisture content ranges from 0 (completely dry) to 1 (completely wet that is fully saturated by water). Water flow in the unsaturated zone is dependent on soil permeability and is unsteady as moisture content is a function of time.

Hydrology, environment engineering, soil mechanics, and water resource engineering are the fields in which the application of one-dimensional groundwater recharge problems is used.

Groundwater recharge has multiple benefits, such as improving agricultural productivity or increasing domestic water for domestic use and providing water for fisheries.

Many researchers have discussed this issue from various perspectives. Klute [1], has studied the flow equation in unsaturated media. Verma [2], has solved the problem of one-dimensional groundwater recharge by spreading using Laplace transformation, whereas Borana et al. [5] have used the Crank Nicolson method. Groundwater recharge through porous media in a vertically downward direction has been discussed by Patel et al. [6]. For finding moisture content in one-dimensional fluid flow through an unsaturated porous medium, Meher et al. [4] used the Adomian decomposition approach. Joshi et al. [8], studied the unsaturated fluid flow through Porous Media. Shah et al. [9] used the q-Homotopy Analysis Method. For solving a one-dimensional vertical groundwater recharge through porous media, Verma et al. [7] employed the singular perturbation approach.

Therefore, the main goal of the present work is to find the solution to one-dimensional groundwater recharge by spreading, using the Hybrid Differential transform and Finite Difference Method, and obtaining the moisture content.

13.2 Research Gap

1. Other numerical methods such as HAM, and HPM need symbolic computations and the introduction of auxiliary operators, auxiliary parameters, and auxiliary functions and perturbation parameters in

the solution procedure which is tedious and time consuming whereas the method HDTFD used in the present work overcomes these limitations.

2. We can improve the approximation by maintaining a balance between the number of intervals in the spatial argument and decreasing the time step suitably which is not possible by the other methods used till now.

3. Due to the limitations of some of the numerical methods, like DTM, RDTM, and TDDTM many researchers have solved the problem using initial conditions only, whereas the present method HDTFD is the combination of the Differential Transform and Finite Difference Method which allows us to use the initial and boundary conditions to obtain the solution.

13.3 Mathematical Formulation

For the flow problem, it is assumed that groundwater recharge occurs over a large basin of such a geological position that there are rigid boundaries on the sides and a thick layer of water table on the bottom. Through spreading, in the unsaturated porous medium water is found to flow vertically downwards. The flow is vertical and goes to a depth of L through an unsaturated porous medium (L is the length of the basin). The Hybrid Differential Transform and the Finite Difference Method (HDTFD) [13] is used to solve Burger's equation [5], which is a one-dimensional nonlinear partial differential equation that describes the phenomenon of groundwater recharge through spreading.

The solution of this equation helps to find the moisture content of the soil.

The equation of continuity for water flow through an unsaturated porous medium is governed by the continuity equation

$$\frac{\partial(\rho_s\theta)}{\partial t} = -\nabla.M \tag{13.1}$$

where, ρ_s= bulk density, θ = moisture content and M= moisture mass flux Water in unsaturated porous medium follows Darcy's law, which is given by

$$v = -k\nabla\emptyset \tag{13.2}$$

where v= moisture volume flux, k =hydraulic conductivity and $\nabla\emptyset$ = gradient of moisture potential.

Relation between mass flux and volume flux of moisture is

$$M = \rho v \tag{13.3}$$

where ρ is the flux density of medium From (13.1), (13.2) and (13.3) using incompressibility of water, we get

$$\frac{\partial(\rho_s \theta)}{\partial t} = -\nabla.(\rho v) \tag{13.4}$$

$$= \nabla.(\rho k \nabla \emptyset)$$

Because in this situation, flow occurs only in the vertical downward direction, which is the z-axis' positive direction.

Equation (13.4) reduces to

$$\rho_s \frac{\partial \theta}{\partial t} = \frac{\partial}{\partial z}\left(\rho k \frac{\partial \psi}{\partial z}\right) - \frac{\partial}{\partial z}(\rho k g) \tag{13.5}$$

where, ψ = pressure potential, g= gravitational constant and

$$\emptyset = \psi - zg \tag{13.6}$$

Rewriting Equation (13.5) as

$$\frac{\partial \theta}{\partial t} = \frac{\partial}{\partial z}\left(D\frac{\partial \theta}{\partial z}\right) - \frac{\rho}{\rho_s}g\frac{\partial}{\partial z}(k) \tag{13.7}$$

where D=$\frac{\rho}{\rho_s} k \frac{\partial \psi}{\partial \theta} = \frac{\rho}{\rho_s} k \alpha$ is called the diffusivity coefficient.

Replacing D_α which is the average value of D for the entire range of moisture content. for D [3] and considering $k \propto \theta^2$ [6]

Therefore $k = k_0 \theta^2$ where k_0 is a constant.

Equation (13.7) reduces to

$$\frac{\partial \theta}{\partial t} + \frac{\rho}{\rho_s}(2gk_0\theta)\frac{\partial \theta}{\partial z} = D_\alpha \frac{\partial^2 \theta}{\partial z^2} \tag{13.8}$$

Writing, $k_1 = \frac{\rho}{\rho_s}(2gk_0)$ equation (13.8) reduces to

$$\frac{\partial \theta}{\partial t} + k_1\theta\frac{\partial \theta}{\partial z} = D_\alpha \frac{\partial^2 \theta}{\partial z^2} \tag{13.9}$$

For simplicity take $k_1 = 1$ and $D_\alpha = 1$.

Choosing new variables,

$Z = \frac{z}{L}$ and $T = \frac{tD_a}{L^2}$, $0 \leq Z \leq 1$ and $0 \leq T \leq 1$

Therefore equation (13.9)becomes,

$$\frac{\partial \theta}{\partial T} + \theta \frac{\partial \theta}{\partial Z} = \frac{\partial^2 \theta}{\partial Z^2} \tag{13.10}$$

Burger's equation (13.10) is the governing nonlinear partial differential equation for determining the moisture content.

As the moisture content of the soil increases with depth Z, the initial moisture content is regarded as an increasing function of Z. As the soil is dry at the top i. e. for Z=0 the value of the moisture content is very small, and at the bottom, as the soil will be fully saturated, the moisture content will be 1.

Therefore, choosing appropriately the initial and boundary conditions, as mentioned below.

Initial Condition

$$\theta(Z, 0) = Ze^{Z-1}, \quad 0 < Z < 1 \tag{13.11}$$

Boundary conditions

$$\theta(0, T) = 0.01 \tag{13.12}$$

$$\theta(1, T) = 1 \quad 0 \leq T \leq 1$$

13.4 Methodology

The Hybrid Differential Transform and Finite Difference Method (HDTFD) is used to solve the non-linear Burger's equation (13.10). The Differential Transform Method (DTM) and the Finite Difference Method (FDM) are combined in this method.

The spatial variables are approximated using the Finite Difference Method, while the time variable is approximated using the Differential Transform Method.

For the solution of linear and nonlinear differential equations in electrical circuit analysis, Zhou [10] proposed the DTM. Chen and Ho [11] improved it by using it to solve partial differential equations.

For the 'T' variable, the differential transform is used, and on the 'Z' variable, the finite difference method is used. The hybrid method is used to solve linear and nonlinear partial differential equations with a few iterations and converges rapidly.

13.5 Hybrid Differential Transform and Finite Difference Method [18]

The Differential Transform of the kth derivative of $u(x, t)$ applied to the 't' variable is given as

$$U(i, k) = \frac{1}{k!}\left[\frac{d^k u(x,t)}{dt^k}\right]_{t=0} \qquad (13.13)$$

$k = 0, 1, 2, \ldots,$ and $i = 0, 1, 2 \ldots$

The inverse Transformation of $U(i, k)$ is given as

$$u(x, t) = \sum_{k=0}^{\infty} U(i, k)t^k \qquad (13.14)$$

where $u(x, t)$ in lower case letters represents the original function and $U(i, k)$ in upper case letters represents transformed function.

$$u(x, t) = \sum_{k=0}^{\infty} \frac{1}{k!}\left[\frac{d^k u(x,t)}{dt^k}\right]_{t=0} t^k \qquad (13.15)$$

where $U(i, k) = U(x_i, k)$, $x_i = ih$, $i = 0, 1, 2, 3, \ldots$.

The finite difference step interval is denoted by h.

13.6 Solution

The Hybrid Differential Transform and Finite Difference Method (HDTFD) and theorems given above are applied to Equation (13.10)

$$\frac{\partial \theta}{\partial T} = \frac{\partial^2 \theta}{\partial Z^2} - \theta\frac{\partial \theta}{\partial Z}$$

Initial conditions:

$$\theta(Z, 0) = Ze^{Z-1}, \quad 0 < Z < 1$$

Boundary conditions:

$$\theta(0, T) = 0.01$$
$$\theta(1, T) = 0 \leq T \leq 1$$

Applying Differential Transforms to the 'T' variable, Finite Difference to the 'Z' variable and by using theorems given in Appendix 1 [15], we get

$$\frac{\partial \theta}{\partial T} = (k+1)\Theta(i, k+1) \qquad (13.16)$$

$$\frac{\partial^2 \theta}{\partial Z^2} = \frac{\Theta\,(i+1.k) - 2\,\Theta\,(i.k) + \Theta(i-1,k)}{h^2} \tag{13.17}$$

$$\theta\frac{\partial \theta}{\partial Z} = \sum_{m=0}^{k} \Theta(i, k-m)\frac{\Theta\,(i+1,m) - \Theta(i-1,m)}{2h} \tag{13.18}$$

where $\theta\,(Z,T)$ is the original function and, $\Theta\,(i, k) = \Theta(Z_i, k)$,$Z_i = ih$, $i = 0, 1, 2, 3, \ldots$ *is* transformed function.

Take differential transform of the boundary conditions:

$$\Theta\,(0, k) = 0.01\delta\,(k)$$
$$= 0.01 \text{ for } k = 0$$
$$= 0 \text{ for } k \neq 0$$
$$\Theta\,(1,k) = \delta(k)$$
$$= 1 \text{ for } k = 0$$
$$= 0 \text{ for } k \neq 0$$

Apply the finite difference method to the boundary conditions and the initial condition.

$$\Theta\,(i, 0) = \Theta(z_i\,, 0) = z_i e^{z_i - 1}, \quad z_i = ih \ , i = 0, 1, 2, 3, \ldots$$

$$\Theta\,(0, k) = .01\delta\,(k)$$
$$= .01 \text{ for } k = 0$$
$$= 0 \text{ for } k \neq 0$$
$$\Theta\,(N, k) = \delta(k)$$
$$= 1 \text{ for } k = 0$$
$$= 0 \text{ for } k \neq 0$$

where N is total no. of spatial segments.

Substituting in equation (13.10) we get, according to the hybrid method, the following recurrence relation

$$(k+1)\,\Theta(i, k+1) = \frac{\Theta\,(i+1.k) - 2\,\Theta\,(i.k) + \Theta(i-1,k)}{h^2} \tag{13.19}$$

$$- \sum_{m=0}^{k} \Theta(i, k-m)\frac{\Theta\,(i+1,m) - \Theta(i-1,m)}{2h}$$

Table 13.1 Moisture content θ Vs Depth Z at fixed time T

Z ↓	T→ 0.001	0.002	0.003	0.004	0.005	0.006	0.007	0.008	0.009	0.01
0	0.01	0.01	0.01	0.01	0.01	0.01	0.01	0.01	0.01	0.01
0.1	0.042369	0.043862	0.045180	0.046357	0.047419	0.048386	0.049275	0.050097	0.050862	0.051573
0.2	0.090850	0.091895	0.092974	0.094065	0.095157	0.096240	0.097310	0.098362	0.099397	0.100421
0.3	0.150024	0.151078	0.152140	0.153212	0.154291	0.155377	0.156466	0.157556	0.158642	0.159717
0.4	0.220673	0.221820	0.222966	0.224112	0.225257	0.226404	0.227551	0.228699	0.229850	0.231007
0.5	0.304505	0.305742	0.306974	0.308204	0.309430	0.310653	0.311873	0.313090	0.314301	0.315506
0.6	0.403502	0.404806	0.406106	0.407400	0.408688	0.409970	0.411246	0.412516	0.413780	0.415039
0.7	0.519917	0.521253	0.522581	0.523899	0.525206	0.526502	0.527785	0.529054	0.530309	0.531548
0.8	0.656305	0.657609	0.658890	0.660147	0.661377	0.662579	0.663755	0.664904	0.666028	0.667128
0.9	0.815526	0.816607	0.817613	0.818554	0.819441	0.820281	0.821079	0.821841	0.822569	0.823268
1	1	1	1	1	1	1	1	1	1	1

where for $k = 0, 1, 2, 3, \ldots$ differential transform coefficients $\Theta(i, 0), \Theta(i, 1), \Theta(i, 2), \ldots$ are obtained. The approximate solutions for various values Z and T are found using the inverse transformation

$$\theta(Z, T) = = \sum_{k=0}^{\infty} \Theta(i, k) T^k \qquad (13.20)$$

$$For, \quad Z_i = 0, i = 0, \qquad \theta(0, T) = \sum_{k=0}^{\infty} \Theta(0, k) T^k$$

$$\theta(0, T) = \Theta(0, 0) + \Theta(0, 1)T + \Theta(0, 2)T^2 + \Theta(0, 3)T^3 + \ldots\ldots = .01$$

where $Z_i = ih$ mesh points for $h = 0.1, \ i = 0, 1, 2, \ldots$

The remaining calculated coefficients from the recursive relation are tabulated and shown in Table 13.1.

13.7 Results and Discussion

The numerical values of the moisture content θ obtained from equation (13.20) for various depths Z at fixed time T=0.001,0.002,0.003,0.004,0.005, 0.006,0.007,0.008,0.009,.01 are acquired using MATLAB and shown in **Table 13.1**

Figure 13.2 shows a graph of moisture content θ versus depth 'Z' for fixed times T=0.001,0.002,0.003,0.004,0.005,0.006,0.007,0.008,0.009, 0.01

Figure 13.2, shows that as the depth Z increases, the moisture content increases for any given time T. This shows that the moisture content distribution in a one-dimensional fluid flow through an unsaturated homogenous porous medium increase steadily with time T > 0. For larger values of 'T',

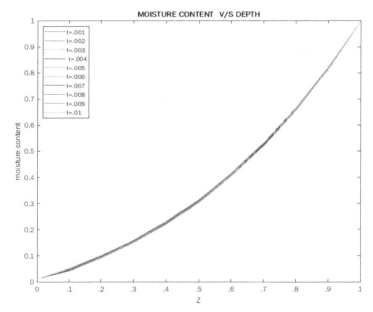

Figure 13.2   Moisture Content Vs. Depth for fixed time T

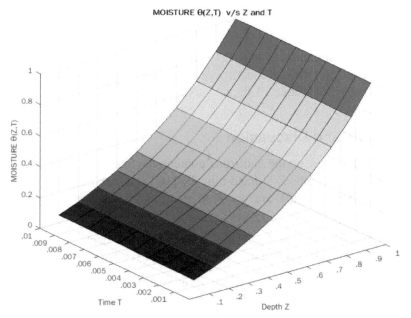

Figure 13.3   Moisture Content Vs. Depth and Time

the solution becomes unstable and this is the limitation of this method, which can be overcome by using the Multi Step Differential Transform Method,

Figure 13.3, is a 3D Plot of moisture content θ versus depth 'Z' and time 'T', which clearly shows how for fixed time 'T' the moisture content θ is increasing as the depth 'Z' is increasing.

13.8 Conclusions

MATLAB is used to obtain numerical solutions and graphs for determining the moisture content θ in one-dimensional groundwater recharge by spreading in unsaturated porous media with respect to depth and time. For a fixed time, T, the initial moisture content θ is lowest at Z=0 and is observed to increase with increasing depth Z.

When compared to other approaches, Hybrid Differential Transform and Finite Difference Method reduces computational difficulties, and all calculations can be performed easily and accurately. The recurrence relation obtained in equation (13.19) has been used to obtain numerical values emphasizing the ability to obtain analytic and numerical solutions by Hybrid Differential Transform and Finite Difference Method which is a blend of the Differential transform Method and Finite Difference Method. Even, the results are found to be reliable and accurate and the series has been found to be convergent.

As a result, we can use this method to solve a wide range of difficult partial differential equations both linear and non-linear without the need for linearization or perturbation. It can be concluded that this method is very powerful and efficient in obtaining numerical solutions for these types of non-linear partial differential equations with initial and boundary conditions.

13.9 Utilities of Research

This research can be utilized to reduce the salinity of the soil due to the increase in moisture content. This technology can also help to reduce the wastage of water and increase groundwater availability for a variety of purposes, including irrigation, home, and industrial usage. It also helps to control soil erosion and flood-like events, as well as supply adequate soil moisture during dry spells or water shortages.

Appendix 1

Theorem 1. *If $g(x,t) = \frac{\partial p}{\partial t}$, then $G(i,k) = (k+1)\, P(i,k+1)$*

Theorem 2. *If $g(x,t) = \frac{\partial^2 p}{\partial t^2}$, then $G(i,k) = (k+1)\,(k+2)\, P(i,k+2)$*

Theorem 3. *If $g(x,t) = xe^{-t}$ .then $G(i,k) = x\frac{(-1)^k}{k!}$*

Theorem 4. *If $g(x,t) = sinx$, then $G(i,k) = sinx$*

Theorem 5. *If $g(x,t) = sint$, then $G(i,k) = sin\left(\frac{\pi k}{2}\right) \frac{1^k}{k!}$*

Theorem 6. *If $g(x,t) = \frac{\partial p}{\partial x}(x,t)$, then*

$$G(i,k) = \frac{P(i+1,m) - P(i-1,m)}{2h}$$

Theorem 7. *If $g(x,t) = p(x,t)\frac{\partial p}{\partial x}(x,t)$, then*

$$G(i,k) = \sum_{m=0}^{k} P(i,k-m)\frac{P(i+1,m) - P(i-1,m)}{2h}$$

References

[1] Klute, A. (1952). A numerical method for solving the flow equation for water in unsaturated materials. Soil Science, 73, 105–116.

[2] Verma, A. P. (1969). The Laplace transform solution of a one-dimensional groundwater recharge by spreading. Annals of Geophysics, 22(13.1), 25–31.

[3] Mehta, M. (1975). A singular perturbation solution of one-dimensional flow in unsaturated porous media with small diffusivity coefficient, Proceedings of the National Conference on Fluid Mechanics and Fluid Power, E1 to E4.

[4] Meher, R., Mehta, M. N., & Meher, S. K. (2010). Adomian decomposition method for moisture content in one dimensional fluid flow through unsaturated porous media. International Journal of Applied Mathematics and Mechanics, 6(13.7), 13–23.

[5] Borana, R. N., et al. (2013). Numerical solution of Burger's equation in a one-dimensional groundwater recharge by spreading using finite difference Method. International Journal of Advance Research in Science and Engineering, 2(13.11), 121–130.

[6] Mehta, M. N., & Patel, T. (2006). A Solution of Burger's Equation Type One Dimensional Ground Water Recharge by Spreading in Porous Media. Journal of the Indian Academy of Mathematics, 28(13.1), 25–32.

[7] Verma, A., & Mishra, S. (1973). A similarity solution of a one-dimensional vertical groundwater recharge through porous media. Revue Roumaine des Sciences Techniques Serie de Mechanique Appliquee, 18(13.2), 345–351.

[8] Joshi, M. S., Desai, N. B., & Mehta, M. N. (2010). One dimensional and unsaturated fluid flow through porous media, Int. J. Appl. Math. and Mech., 6(13.18), 66–79.

[9] Shah, K., & Singh, T. (2016). Solution of Burger's Equation in a One-Dimensional Groundwater Recharge by Spreading Using q- Homotopy Analysis Method. European Journal of Pure and Applied Mathematics, 9(13.1), 114–124.

[10] Zhou, J. K. (1986). Differential Transformation and Its Applications for Electrical Circuits. Huazhong University Press, Wuhan.

[11] Chen, C. O. K., & Ho, S. H. (1999). Solving partial differential equations by two-dimensional differential transform method. Applied Mathematics and computation, 106(2-3), 171–179.

[12] Arslan, D. (2020). The Numerical Study of a Hybrid Method for Solving Telegraph Equation. Applied Mathematics and Nonlinear Sciences, 5(13.1), 293–302.

[13] Süngü, İ.Ç., & Demir, H. (2012). Application of the Hybrid Differential Transform Method to the Nonlinear Equations. Applied Mathematics,3(13.3), 246–250.

[14] Ayaz, F. (2003). On the two-dimensional differential transform method. Applied Mathematics and Computation, 143(2–3), 361–374.

[15] Arslan, D. (2020). The Comparison Study of Hybrid Method with RDTM for Solving Rosenau-Hyman Equation. Applied Mathematics and Nonlinear Sciences, 5(13.1), 267–274.

[16] Patel, M. A., & Desai, N. B. (2019). An approximate analytical solution of one-dimensional groundwater recharge by spreading. Journal of Applied and Engineering Mathematics, 9(13.4), 838–850.

[17] Parikh, A., & Mehta, M. (2011). Transcendental Solution of Fokker-Planck Equation of Vertical Ground Water Recharge in Unsaturated Homogeneous Porous Media. International Journal of Engineering Research and Applications, 1(13.4), 1904–1911.

[18] Hatami, M., et. al. (2017) Differential Transformation Method for Mechanical Engineering Problems. Academic Press.

14

Numerical Solution of Physiological Thermoregulatory Disturbances in Cold Environment

Akshara Makrariya[1*], Rahul Makrariya[2], and Manisha jain[1]

[1]School of Advanced Science- Mathematics, VIT University Bhopal
[2]Sagar Institute of Research and Technology, Bhopal-MP, India
E-mail: aksharahul@gmail.com
*Corresponding Author

Abstract

The interest in the human body's physiological capacity to adapt to extreme heat and cold conditions has increased enormously in the last few decades because of global warming and the consequent changing temperatures. The human body has multiple thermoregulatory mechanisms to counter the external extreme temperatures whose main objective is to keep temperature homeostasis within normal values. As exposure time to these stressful conditions increases and the external temperature becomes even more extreme, the body systems start adapting to its environment progressively. All of the adaptations, at the beginning of the exposure somewhat irrelevant, but may become very important since they can affect all of the body systems in a negative manner and finally, compromise life. Clinically, all of these responses and adaptations are manifested through clinical signs and symptoms. The purpose of this paper is to discuss how human physiology adapts during extreme heat and cold conditions to maintain temperature homeostasis and its consequences when this cannot be achieved.

Key Words: Micro-circulatory apparatus, Cold Environment, Intracellular Fluid Crystallises.

14.1 Introduction

The Heat transfer process in the human physiological system is highly complex due to several factors such as blood perfusion and distribution, intra-cellular flow of lipids and water, cellular metabolic reaction with heat generation and absorption, hanging patterns of fats and nucleus, and hypothalamic control of flow process based on thermal receptors. Apart from this, there are several other associated physical activities and physiological disturbances which cause erraticism and abnormality in the thermal engineering aspect of the whole system.

The internal parts of the human body are maintained in a natural way within a narrow range of prescribed temperatures. However, the outer layer including the skin undergoes various thermal situations depending on the atmospheric change and nervous stimulation. The skin is the principal site of heat loss during exposure to a cold environment or heat gain during exposure to a hot environment.

Extensive cooling of skin parts causes physiological damage due to following reasons:

1. The intracellular fluid crystallizes at low temperature followed by water crystallization a subfreezing atmosphere in direct control with the outer surface. This causes ruptures in nerve fibers and in-vivo cells. The porosity of the region increases due to this process and the amount of water evaporation rises.

2. The micro-circulatory apparatus is disturbed greatly at low temperatures. The red blood cells are sick with each other and the capillaries are blocked. Therefore, the supply of warm blood from the internal part is reduced and the heating network becomes less efficient. Another negative effect of low temperature is on the capillary walls which are made of endothelium cells. The result of the effect is the transudation of fluid edema.

3. Physical and chemical effect of cold on the tissue is also significant due to the fall in the thermal conductivity and rate of cell chemical reaction. The latter is due to the disturbance in nutrient concentration inside. There are several stages of damage due to cold. In some cases, these damages are reversible and curable while in prolonged and excessive cooling, the changes are irreversible and necrotic. The mathematical and computational study of thermo-regulation in the skin, which plays the main role in balancing the heat transport process between the outer and internal environment has already been carried out by Saxena and

his co-workers. Arya and Saxena(1); Saxena(6), Saxena and Bindra(8) for various configurations and situations. They have modified the tissue heat transfer model by Perl(2.3) and employed it to compute temperature profiles in human dermal layers. The heat diffusion equation due to Perl was later studied by cooper and Trezek(4) for brain tissue with a negligible effect on blood flow and metabolic heat generation. The same model was used and applied for steady and no steady-state cases by Chao, Eisley, and Yang(5). Perl's model was used later by Saxena and Pardasani(9) for cancerous tumor problems situated in the peripheral region of the human body. The problems including tissue damage based on temperature rise beyond threshold value are also studied by Saxena(7). The recent movement in this direction is to study and analyze the effect of freezing atmosphere on tissue Saxena and Varma(10).

In view of the above here two models have been developed to study temperature distribution in spherical human organs for a one-dimensional steady-state case [11, 12].

The first model is developed using the analytical method and a solution is obtained in terms of Bessel's function [13]. The second model is developed using the finite element method. The numerical results are obtained to study temperature distribution in the skin and subcutaneous tissues under normal, physiological, and environmental conditions. The physiology of skin required to develop the model is given in the next section [11, 14].

Based on the physiological background and mathematical background given in previous research, a model for heat flow in a spherical human organ is presented here for a one-dimensional steady-state case. The analytical method has been employed for finding the solution to the problem under normal physiological and environmental conditions [15, 16].

Kumar et al., [29] also gave the exact solution to the Penne's bioheat equation in one-dimensional multilayer spherical geometry [26]. Prasad et al. developed an inverse method to optimize the heating condition during a hyperthermia treatment [27].

14.2 Material and Methods

Conduction of heat takes place through tissue and other constituents including almost stationary blood and other fluids. However, the physical parameters like thermal conductivity, density, and specific heat do not have fixed values and change from one point to another and also with time (in transient cases).

The mathematical form of this conduction process is

$$\rho_i c_i \left(\frac{\partial T}{\partial t}\right)\Big|_C = \frac{\partial}{\partial x}[K_i(x,t)]\frac{\partial T}{\partial x} \tag{14.1}$$

Here "i" denotes the sub-region of the skin layer in which the conduction process is defined. Also, the perfusion of blood contributes dominantly to the heating of the whole body through the skin capillary bed. This perfusion can be put in the following form

$$\rho_i c_i \left(\frac{\partial T}{\partial t}\right)\Big|_P = \rho_b c_b \left(\varphi_A T_A - \varphi_V T_V\right) \tag{14.2}$$

where T, T_A, T_V are local tissue, arteriole blood and venular blood temperatures respectively. ρ_b and c_b are density and specific heat of blood.

The domain structure reveals that each sub-region should be distinctly represented in any mathematical thermal model at low atmospheric temperature. In the case when the whole region is considered to be homogeneous the model in variational form can be represented as

$$I = \frac{1}{2}\int_{a_i}^{a_{i-1}}\left[K^{(i)}\left(\frac{\partial T^{(i)}}{\partial x}\right)^2 + P_b C_b T^{(i)^2}(\lambda_A - \lambda_V)\right.$$
$$\left. - 2S^{(i)}T^{(i)} + \rho c\frac{\partial T^{(i)^2}}{\partial t}\right]dx \tag{14.3}$$

With the following boundary and initial conditions

$$K\frac{\partial T}{\partial x}\Big|_{X=a_0} = h\left(T_a - T\right) + LE \tag{14.4}$$

and

$$T(x,o) = -5.8 + px \tag{14.5}$$

where p is a known constant. Here $K, p,$ and c are defined to be thermal conductivity, density, and specific heat respectively of the tissue. h is a heat transfer coefficient. The rate of heat generation, sweat evaporation, and latent heat of evaporation are denoted by S, E and L. λ_A and λ_V are the fraction of the internal blood temperature entry and going out of the tissue.

This form has to be redefined in each physiologically distinct part of the dermal region under the full or partial effect of cold injury. In the broad portion wise model, the variational form(3) is redefined in the region as Ic, Ig, lp, Ir and Is respectively for Stratum Corncum $(a_0, < x < a_1)$, Stratum Germinativum $(a_1 < x < a_2)$, Papillary region $(a_2 < x < a_3)$, and subdermal part $(a_3 < x < a_4)$.

In Stratum Germinativum part the metabolic rate will fall rapidly and will vanish as soon as the injury sets in. However, an additional term might have to be incorporated in case the water transport at the surface is excessive. In Papillary region the pattern of blood perfusion rate i.e. the value of λ_A and λ_V will depend on the stage of the "hunting reaction". For example, $\lambda_A = \lambda_V$, when the blood flow is negligible and $\lambda_V = 0$. then the quantity of blood in this part is maximum. However, the rate of metabolic activity will continuously fall due to less water available to the active cells.

The above variational form in each region have to be assembled by using the relation

$$I = \sum I_N \tag{14.6}$$

where $N = c, g, p, r, s$,

Let $T'(x)$ $(r = 1, 2, 3, 4, 5)$ represent the linear variation in $T(x)$ in all the five region in which SST region has been subdivided. Then evaluating the variational integrals for all the five parts and minimizing the same with respect to $T0$, $T1$, $T2$, $T3$ and $T4$ since $T5$ is known, we get the following equations.

$$\frac{1}{6} \rho c \left[2a_1 T_0' + a_1 T_1'\right] + \left(\frac{K_1}{a_1} + h\right) T_0 - \frac{K_1}{a_1} T_1 = hT_a - LE$$

$$\frac{1}{6} \rho c [a_1 T_0' + 2a_2 T_1' + (a_2 - a_1) T_2' + N_1 T_1 + L_1 T_2 = W_1 - M_1 T_0$$

$$\frac{1}{6} \rho c \left[2(a_3 - a_1) T_2' + (a_2 - a_1) T_1' + (a_3 - a_2) T_3'\right] =$$
$$W_2 + SM_2 T_1 + N_2 T_2 + L_2 T_3 \tag{14.7}$$

$$\frac{1}{6} \rho c \left[2(a_4 - a_2) T_3' + (a_3 - a_2) T_2' + (a_4 - a_3) T_4'\right] = W_3 + N_3 T_2 + L_3 T_3 + R_3 T_4$$

$$\frac{1}{6} \rho c \left[2(a_4 - a_3) T_3' + (a_5 - a_3) T_4'\right] + L_4 T_3 + R_4 T_4 = W_4$$

With the help of Euler's method, the equations (14.7) are transformed into system of algebraic equations, as

$$\frac{1}{6} \rho c [2a_1 T_0 + a_1 T_1] = A_{11} T_{00} + A_{12} T_{01} + (hT_a - LE)$$

$$\frac{1}{6}\,\rho c\,[a_1 T_0 + 2a_2 + (a_2 - a_1)T_2] = A_{22}T_{01} + A_{23}T_{02} + W_1 \qquad (14.8)$$

$$\frac{1}{6}\,\rho c\,[(a_2 - a_1)T_1 + 2\,(a_3 - a_1)\,T_2 + (a_3 - a_2)T_3] = A_{32}T_{01} + A_{33}T_{02} + A_{34}T_{03}$$
$$+ W_2$$

$$\frac{1}{6}\,\rho c\,[(a_3 - a_2)T_2 + 2\,(a_4 - a_2)\,T_3 + (a_4 - a_3)T_4] = A_{43}T_{02} + A_{44}T_{03} + A_{45}T_{04}$$
$$+ W_3$$

$$\frac{1}{6}\,\rho c\,[(a_4 - a_3)T_3 + 2\,(a_5 - a_3)\,T_4] = A_{54}T_{03} + A_{55}T_{04} + W_4$$

The equations (14.8) are simplified for $T0, T1, T2, T3$ and $T4$ using the initial condition (14.5) at each of the interface.

14.3 Result

To solve equations (14.8) we make use of the following value of the physical parameters.

T. $= -6\,°C$, p $= 1.05$ grn/cm^2, c $= 0.83$ Cal/ gm,
   E $= 0.24$ x 10-3 grn/cm^2 / min
h $= 0.01$ cal/ cm^{2}min C, L $=579$ cal/gm , $\lambda_A = 0.6, \lambda_V = 0.5$,
   $T_b = 37\,°C$
$K_1 = 0.015$ cal / cm/min $°C$, $K_2 = K_1$, $K_3 = 0.0\ 225$ cal/cm/min $°C$,
   $K_4 = K_3$,
$K_5 = 0.03$ cal / cm/min $°C$
$m_1 = 0$, $m_2 = 0$, $m_3 = 0.0009$ cal / cm^3 / min $°C$,, $m_4 = 0.0012$ cal /
   cm^3 / min $°C$,
$m_5 = 0.0018$ cal / cm^3 / min $°C$,
$S_1 = 0$, $S_2 = 0.004$ cal / cm^3 / sec., $S_1 = 0.0178$, cal / cm^3/ sec.
$S_4 = 0.0178$ cal / cm^3 / sec. $S_5 = 0.0357$ cal / cm^3 / sec.
The following set of thicknesses has been used (all in cms.)
$0.15, 0.31, 0.45, 0.6, 0.75$.

The temperature distribution in each of the region as a function of time thus obtained are given below:

$$T_0 = -6.99\ \delta\,t - 5.79;$$
$$T_1 = \ 0.66\ \delta\,t + 2.75\ ;$$

$$T_2 = 4.28\,\delta\,t + 11.30$$
$$T_3 = 3.13\,\delta\,t + 19.84\;;$$
$$T_4 = 6.89\,\delta\,t + 28.42$$

14.4 Discussion

Temperature distribution for a single set of values of layer thickness has been obtained. The initial temperature is assumed to vary linearly with position according as $T_i(x) = T_0 + PX_0$

$i=$ 0, 1, 2, 3, 4. The nodal temperatures are calculated using Euler's method. The skin surface temperature Too has been taken as a guiding parameter, which in turn is assumed to be constant and approximately equal to the environmental temperature. This may also be taken as a function of time. Such variations are common to observe when the thermal conductivity, rates of blood mass flow, and metabolic heat generation have been assigned different values for atmospheric temperatures. Evaporation rate has been assigned a small value due to an increase in the porosity of the region.

In order to obtain the interface temperature at any time t the use of Euler's method has been made. The analysis of the results thus obtained shows unstable temperature distribution after a certain period of time. One thing we noticed is that this method is helpful only for a small interval of time and for long time analysis we have to go for some other numerical method.

References

[1] Patil, H. M., R.J.T.S. Maniyeri, and E. Progress, Finite difference method based analysis of bio-heat transfer in human breast cyst. 2019. 10: p. 42-47.
[2] Makrariya, Akshara, and N. Adlakha. "Thermal stress due to tumor in periphery of human breast."J Biosci Bioeng 2 (2015): 50-59.
[3] Makrariya, Akshara, and K. R. Pardasani. "Finite Element Model To Study The Thermal Effect Of Cyst And Malignant Tumor In Women's Breast During Menstrual Cycle Under Cold Environment." Advances and Applications in Mathematical Sciences 18.1 (2018): 29-43.
[4] Khanday, M. A., A.J.J.o.M.i.M. Rafiq, and Biology, Numerical estimation of drug diffusion at dermal regions of human body in transdermal drug delivery system. 2016. 16(3): p. 1650022.

[5] Gonzalez, F.J.J.B.S. and Imaging, Theoretical and clinical aspects of the use of thermography in non-invasive medical diagnosis. 2016. 5(4): p. 347-358.

[6] Choi, J. S., *et al.*, Toxicological effects of irregularly shaped and spherical microplastics in a marine teleost, the sheepshead minnow (Cyprinodon variegatus). 2018. 129(1): p. 231-240.

[7] Aziz, Rabia, C. Verma, and N. Srivastava. "A weighted-SNR feature selection from independent component subspace for nb classification of microarray data." Int J Adv Biotec Res 6 (2015): 245-255.

[8] Musheer, R. Aziz, C. K. Verma, and N. Srivastava. "Novel machine learning approach for classification of high-dimensional microarray data."Soft Computing 23.24 (2019): 13409-13421.

[9] Aziz, R., C. K. Verma, and N. Srivastava, Artificial neural network classification of high dimensional data with novel optimization approach of dimension reduction. Annals of Data Science, 2018. 5(4): p. 615-635.

[10] Aziz, R., N. Srivastava, and C. K. Verma. "T-independent component analysis for svm classification of dna-microarray data." International Journal of Bioinformatics Research, ISSN (2015): 0975-3087.

[11] Makrariya, A. and Neeru or N. Adlakha, Quantitative study of thermal disturbances due to nonuniformly perfused tumors in peripheral regions of women's breast. 2017. 16: p. 1176935117700894.

[12] Aziz, R., Verma, C. K., Jha, M., and Srivastava, N. (2017). Artificial neural network classification of microarray data using new hybrid gene selection method. International Journal of Data Mining and Bioinformatics, 17(1), 42-65.

[13] Zanoni, M., *et al.*, 3D tumor spheroid models for in vitro therapeutic screening: a systematic approach to enhance the biological relevance of data obtained. 2016. 6(1): p. 1-11.

[14] Farmani, S., M. Ghaeini-Hessaroeyeh, and S.H.J.I.J.f.N.M.i.F. Javaran, The improvement of numerical modeling in the solution of incompressible viscous flow problems using finite element method based on spherical Hankel shape functions. 2018. 87(2): p. 70-89.

[15] Makrariya, A., K. R. Pardasani, Modelling, Numerical simulation of thermal changes in tissues of woman's breast during menstrual cycle in different stages of its development. 2019. 14(4): p. 348-359.

[16] Makrariya, A., N. Adlakha, and S. K. Shandilya. 3D Spherical—Thermal Model of Female Breast in Stages of Its Development and Different Environmental Conditions. in Mathematical Modeling, Computational Intelligence Techniques and Renewable Energy:

Proceedings of the First International Conference, MMCITRE 2020. 2021. Springer.

[17] Salari, M., *et al.*, Numerical analysis of turbulent/transitional natural convection in trapezoidal enclosures. 2017.

[18] Makrariya, A., K. R. Pardasani, Numerical study of the effect of non-uniformly perfused tumor on heat transfer in women's breast during menstrual cycle under cold environment. Network Modeling Analysis in Health Informatics and Bioinformatics (2019) 8:9.

[19] Chao, K. and W.J.B.-M.s. Yang, ASME, Response of skin and tissue temperature in sauna and steam baths. 1975: p. 69-71.

[20] Makrariya, A. and Neeru or N. Adlakha, Thermographic pattern's in women's breast due to uniformly perfused tumors and menstrual cycle. Commun. Math. Biol. Neurosci. 2019 (2019): Article ID. 14.

[21] Pandey, H.R.J.J.o.A.C.o.E. and Management, A One-Dimensional Bio-Heat Transfer Equation with Galerkin FEM in Cylindrical Living Tissue. 2015. 1: p. 45-50.

[22] Roy, K. S., *et al.*, Polymeric sorbent with controlled surface polarity: an alternate for solid-phase extraction of nerve agents and their markers from organic matrix. 2018. 90(11): p. 7025-7032.

[23] Khanday, M., *et al.*, Numerical estimation of the fluid distribution pattern in human dermal regions with heterogeneous metabolic fluid generation. 2015. 15(01): p. 1550001.

[24] Flicek, K., *et al.*, Correlation of radiologic with surgical peritoneal cancer index scores in patients with pseudomyxoma peritonei and peritoneal carcinomatosis: how well can we predict resectability? 2016. 20(2): p. 307-312.

[25] Takahashi, M., *et al.*, Ag/FeCo/Ag core/shell/shell magnetic nanoparticles with plasmonic imaging capability. 2015. 31(7): p. 2228-2236.

[26] Kumar, D., *et al.*, A study on DPL model of heat transfer in bi-layer tissues during MFH treatment. 2016. 75: p. 160-172.

[27] Prasad, B., *et al.*, Effect of tumor properties on energy absorption, temperature mapping, and thermal dose in 13.56-MHz radiofrequency hyperthermia. 2018. 74: p. 281-289.

[28] Makrariya, A. and Neeru or N. Adlakha, Two-dimensional finite element model to study temperature distribution in peripheral regions of extended spherical human organs involving uniformly perfused tumors. International Journal of Biomathematics. Vol. 08, No. 06, 1550074 (2015).

[29] Mignone, A.J.J.o.C.P., High-order conservative reconstruction schemes for finite volume methods in cylindrical and spherical coordinates. 2014. 270: p. 784-814.

[30] Cho, Y. and S.J.I.U.M.J. Lee, Strichartz estimates in spherical coordinates. 2013: p. 991-1020.

[31] Bezerra, L., *et al.*, An empirical correlation to estimate thermal properties of the breast and of the breast nodule using thermographic images and optimization techniques. 2020. 149: p. 119215.

[32] Grombein, T., K. Seitz, and B.J.J.o.G. Heck, Optimized formulas for the gravitational field of a tesseroid. 2013. 87(7): p. 645-660.

[33] Delouei, A. A., *et al.*, A closed-form solution for axisymmetric conduction in a finite functionally graded cylinder. 2019. 108: p. 104280.

[34] Zhang, X., *et al.*, A closed-form solution for the horizontally aligned thermal-porous spheroidal inclusion in a half-space and its applications in geothermal reservoirs. 2019. 122: p. 15-24.

15

Mathematical Modelling of Transient Heat Conduction in Biological System by Finite Element Method and Coding in MATLAB

Manisha Jain and Akshara Makrariya

Vellore Institute of Technology, Bhopal University
E-mail: mujain31@gmail.com; Manisha.jain@vitbhopal.ac.in

Abstract

The numerical displaying of biological structures needs a strategy that can give the appropriate procedure. The vast majority of the numerical models of biological structures are nonlinear and transient in nature and manage the fractional differential conditions. The mathematical techniques are the better decision for such models. Heat move in the natural structure is one of the significant spaces of use of limit issues. Such circumstances are normally experienced in heat moving through composite layers of skin exposed to internal and external temperatures that are protected toward one side. The Finite Element Method (FEM) is the most generally utilized mathematical technique to settle Partial Differential Equations. This paper manages the arrangement of the one-layered heat condition utilizing the Finite Element Method and coding in MATLAB.

Keywords: Partial Differential Equation, Transient Equation, Biological System, Finite Element Method, Boundary Problems, MATLAB.

15.1 Introduction

This article deals with the mathematical modeling of transient heat conductions problems which has various applications in many practical engineering areas [1–4]. The temperature of a body and associated parameters vary with time as well as position. Under steady conditions, the temperature of a body at any point does not change with time during heat conduction [5–10]. In such cases, heat conduction is a function of position only. The temperatures of a body at any point vary with time and position in one dimensional and multidimensional system in heat conduction under unsteady state conditions. Heat distribution problems require the consideration of the complete time history of the temperature variation.

It is very difficult to find the analytical solution to the irregular boundaries of the heat transfer region. The Numerical Method is the best choice to solve such complicated heat transfer problems with high performance computing analysis numerically. The standard finite difference [11], the finite volume method [12–13], and finite element methods have been developed to solve the heat equations. The advantage of the FEM method is that the general-purpose computer program can be developed easily to analyze complicated heat transfer problems. FEM is useful to handle regions with irregular shapes and irregular boundary conditions. The approximate solution of the time-dependent Partial Differential equation can be originated using the Finite Element Method (FEM). The governed differential equation (steady-state problems) or the Partial Differential Equations will be solved numerically using standard techniques i.e Runge –Kutta Method, Euler's method, or Crank Nicolson Method, etc.

A numerically stable equation is required to approximate the partial differential equation. Numerical algorithms require numerical stability. Numerical stability depends on the situation and the accuracy of the algorithm. Inaccuracies in the input data and intermediate calculations do not accumulate and cause the resulting output to be meaningless [14]. The Finite Element Method is an accurate tool for solving partial differential equations for complex domains.

MATLAB is one of the best software for handling such complex modeling. Programing of the Finite Element Method can be done in MATLAB to optimize the results of Biological Problems.

15.2 General Procedure of Finite Element Method

The Finite Element Method is a numerical analysis technique for obtaining the approximate solution of physical problems governed by a differential equation. The most distinctive feature of the FEM is to divide the entire domain into sub domains. Each sub-domain is called the finite element; thus, the name of this method is the finite element method. The Finite element allows applying various initial or boundary values to distinct layers and their elements also.

The FEM method is widely described in the literature and used for solving, heat transfer problems [1–2, [5], [8], 15–18 behavioral analysis of various structures [1–2, 6–7] and to study fluid flow phenomena [1, 4]. This method utilizes an integral formulation to generate a system of algebraic equations. Continuous smooth functions can be applied for approximating the unknown quantities.

Following are some steps that are very important for solving heat equations using FEM and solution in MATLAB

15.3 Process of Finite Element Method

Step-1: Discretization of the Domain-based on the following criteria [19]

- Element Types - Linear or Non-Linear
- Element Size - (One dimensional, Two Dimension, and Three Dimensional)
- Location of Nodes
- Number of Elements
- Node Numbering Scheme
- Automatic Mesh Generation: By Using MATLAB
- Selection of approximate interpolation function
- Shape Function
- Characteristics of Shape Functions
- Establishing the Coordinate Systems
- Natural Coordinate System
- Classical Interpolation Functions - Lagrange Polynomials (Lagrange Interpolation Functions) and Hermitian Polynomials
- Matrix Derivation and Element Vectors

The characteristic matrices and characteristic vectors are also known as vectors of nodal actions of finite elements can be derived by using any of the following approaches:

- Direct Approach
- Variational Approach: Weak form
- Method of Weighted Residual Approach: Weak form
- Solution of Equilibrium Problems using Variational: (Rayleigh-Ritz) Method

- Assembly of The Global Matrix Equation
- Incorporation of Boundary Conditions
- Numerical Solution of Finite Element Equations
- Interpretation of the result

15.3.1 Definition of the Problem and its Domain: One Dimensional Thermal Equation of Biological System

Heat travels from the body core to the body surface through the skin in the biological system. The skin is the most important and the largest organ of the integumentary system made up of multiple layers. It is the first line of defense, which covers and guards the underlying muscles, bones, ligaments, and internal organs. Skin is the interface with the environment, and plays an important role in protecting against infectious [20]. The skin is anatomically separated into three sub-layers viz. epidermis, dermis, and hypodermis or subcutaneous layer.

The study of body temperature is very important in the analysis and prediction of any type of disease because a small disturbance in the mechanism will cause lots of complications in the natural processing of the body. Biological processes are affected by body core temperature (T_b). Core temperature is the balance between heat gain produced by cell metabolism and heat loss by various mechanisms. Apart from the core temperature, the measurement of the surface temperature of the human body is also important [21].

The heat and mass distribution equation in biological systems is given by Perl [22] in terms of the partial differential equation.

$$Div(K\,grad\,T\,) + m_b c_b\,(T_b - T) + S = \rho c \frac{\partial T}{\partial t} \qquad (15.1)$$

Here the effect of metabolic heat generation and blood mass flow is given by the terms S and $m_b c_b (T_b\text{-}T)$ respectively. T_b, K, ρ, c, m_b, and c_b are body core temperature, thermal conductivity, density, and specific heat of tissue; blood mass flow rate and specific heat of blood respectively.

Suppose the outer surface of the region is exposed to the environment. Therefore, net heat flux is dissipated from x_n boundary only. Heat loss from the outer surface due to conduction, convection, radiation, and evaporation is calculated by following boundary conditions as

$$-K \frac{\partial T}{\partial x}\bigg|_{x=x_n} = h(T - T_a) + LE \quad for \;\; t > 0 \qquad (15.2)$$

Here, h, T_a, L, E, and $\partial T/\partial n$, are heat transfer coefficient, atmospheric temperature, the latent heat, rate of evaporation, and the partial derivatives of T along the normal to the skin surface respectively.

As the body core temperature is maintained at uniform temperature Tb, therefore inner boundary condition is prescribed by the Dirichlet boundary condition as

$$T(x,t)|_{x=x_o} = T_b \text{ for } t \geq 0, \ T_b = 37\,^{\circ}\text{C} \qquad (15.3)$$

Assuming initially at the time t = 0 the outer surface of the skin is assumed to be insulated hence the initial condition is given by

$$T(x,0) = T_b \qquad (15.4)$$

Here the effect of metabolic heat generation and blood mass flow is given by the terms S and $m_b c_b$(Tb-T) respectively. T_b, K, ρ, c, m_b, and c_b are body core temperature, thermal conductivity, density, and specific heat of tissue; blood mass flow rate and specific heat of blood respectively.

15.4 Steps Involved in Finite Element Process

Step-1: Discretization of the Domain: One Dimensional Simplex Element

Model 1: One Dimensional Linear Interpolation

Divide the skin surface into several finite elements i.e a one-dimensional element of length l with two nodes, one at each end say i and j and the nodal values of the field variables T as T_i and j. as shown in Figure 15.2.

Step-2: Assumption of a suitable form of variation in T

i. For Linear Element

Assume a linear temperature inside any element "e" as ε_1 and ε_2 for each element e are determined as follows:

$$T^{(e)}(x,t) = \varepsilon_1 + \varepsilon_2 x = [N(x)]\vec{T}^{(e)}$$
$$T(x) = T_i \ at \ x = x_i \ and \ T(x) = T_j \ at \ x = x_j \qquad (15.5)$$

Equation (15.5) can be obtained

$$T_i(x,t) = \varepsilon_1 + \varepsilon_2 x_i \ and \ T_j(x,t) = \varepsilon_1 + \varepsilon_2 x_j$$

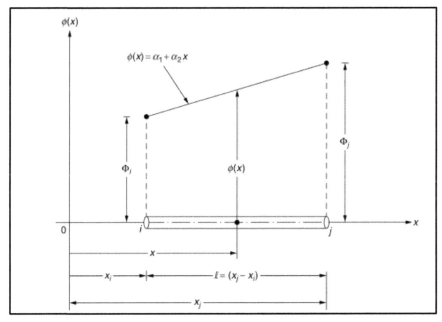

Figure 15.1   One dimensional simplex elements for interpolation

The solution of these two equations gives:

$$T_i^{(e)}(x,t) = T_1 = \frac{T_i x_j - T_j x_i}{l^{(e)}} \ and \ T_j(x,t) = T_2 = \frac{T_j - T_i}{l^{(e)}}$$

$$\Rightarrow T^{(e)}(x,t) = \left(\frac{T_i x_j - T_j x_i}{l^{(e)}}\right) + \left(\frac{T_j - T_i}{l^{(e)}}\right) x \ by \ (5)$$

$$T^{(e)}(x,t) = [N_i(x)]\, T_i + [N_j(x)]\, T_j = [N(x)]\, \overrightarrow{T}^{(e)}$$
$$[N(x)] = [N_i(x) N_j(j)]$$

where

$$N_i(x) = N_1(x) = \frac{x_j - x}{l^{(e)}} = 1 - \frac{x}{l^e}$$

$$N_j(x) = N_2(x) = \frac{x - x_i}{l^{(e)}} = \frac{x}{l^e}$$

$$\bar{T}^e = \begin{bmatrix} T_i \\ T_j \end{bmatrix}, (l = x_j - x_i), = vector \ of \ nodal \ unknown \ of \ element \ 'e'$$

15.5 Model-2: One Dimensional Quadratic Interpolation Model

15.5.1 Assumption of a Suitable form of Variation in T for Quadratic Element

The region consists of sufficiently large numbers of elements having a very small size. Each element consists of 3 nodes namely i, j, and k. T with quadratic shape function is assumed as:

$$T(x, t) = \varepsilon_1 + \varepsilon_2 x + \varepsilon_3 x^2 \qquad (15.6)$$

where ε_1, ε_2 and ε_3 are three unknown constants in Eq. (15.6), the element is assumed to have three degrees of freedom, one at each of the ends and one at the middle point as shown in Figure (15.3) at each node [18]

$$T = T_i \ at \ x = 0; T = T_j \ at \ x = \frac{1}{2}; T = T_k \ at \ x = l \qquad (15.7)$$

where l is the length of the element e. The values of ε_1, ε_2 and ε_3 can be calculated as

$$\varepsilon_1 = T_i; \varepsilon_2 = \frac{(4T_j - 3T_i - T_k)}{l}; \varepsilon_3 = \frac{(2T_i - 2T_j - T_k)}{l^2}$$

With the help of eq. (15.7), eq. (15.6) can be expressed after rearrangement as:

$$T(x, t) = [N(x)]\bar{T}^e$$

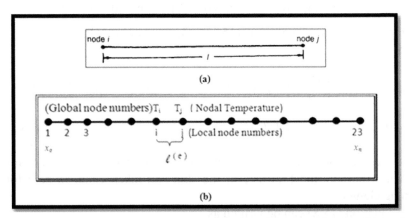

(a)

(b)

Figure 15.2 (a) 1-D Linear Element (b) Discretization of domain using One Dimensional Linear Element

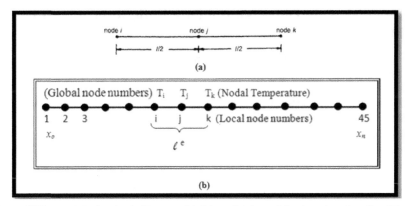

Figure 15.3 (a) 1-D Quadratic Element (b) Discretization of the domain using One Dimensional Quadratic Element

where

$$[N(x)] = [N_i(x)N_j(x)N_k(x)], \bar{T}^e = [T_iT_jT_k]'$$

$$N_i(x) = \left(1 - \frac{2x}{l^{(e)}}\right)\left(1 - \frac{x}{l^{(e)}}\right), N_j(x) = \frac{4x}{l^{(e)}}\left(1 - \frac{x}{l^{(e)}}\right),$$

$$N_k(x) = -\frac{x}{l^{(e)}}\left(1 - \frac{2x}{l^{(e)}}\right);$$

$$l^{(e)} = (x_k - x_i)$$

Step-3 : Express the function I as a sum of E Elemental Identities I^e

The vibrational form of equation (15.1) using equations ((15.2)–(15.4)) is given by

$$I^e = \frac{1}{2}\int_0^l [K^e\left(\frac{\partial T^e}{\partial x}\right)^2 + M^e(T_b - T^e)^2 - 2S^eT^e + \rho c\frac{\partial (T^e)^2}{\partial t}]dx$$

$$+ \frac{1}{2}[h(T^e - T_a)^2 + 2LET^e] \tag{15.8}$$

Here second integral of the equation (15.8) is valid for the elements e adjoining the outermost surface of the skin and taken equal to zero for remaining elements.

The equation (15.8) can be written as

$$I^e = I_K^e + I_M^e - I_S^e + I_\rho^e + I_{\delta1}^e + I_{\delta2}^e \tag{15.9}$$

where,

$$I_K^e = \frac{1}{2} \int_e K^e \left(\frac{\partial T^e}{\partial x}\right)^2 dx, \; I_M^e = \frac{1}{2} \int_e \left[M^e \left(T_b - T^e\right)^2\right] dx$$

$$I_s^e = \int_e [S^e T^e]\, dx, \; I_\rho^e = \frac{1}{2} \int_e \rho c \frac{\partial (T^e)^2}{\partial t}\, dx$$

$$I_{\delta 1}^e = \frac{1}{2} \left[h \left(T^e - T_a\right)^2\right], \; I_{\delta 2}^e = LET^e$$

After minimizing equation (15.9) with respect to T_i, The matrix form of eq. as follows

$$\frac{dI^e}{d\bar{T}^e} = [A_1^e]\,[\bar{T}^e] + [A_2^e]\,[\bar{T}^e] - [A_3^e] + [A_4^e]\left\{\frac{\partial \bar{T}^e}{\partial t}\right\} + [A_5^e]\,[\bar{T}^e] + [A_6^e]$$

$$(15.10)$$

$$[A_1^e]_{p \times p} = \int_e K^e [B^e]'\, [B^e]\, dx, \; [A_2^e]_{p \times p} = \int_e M^e [N^e]'\, [N^e]\, dx$$

$$[A_3^e]_{p \times 1} = \int_e (M^e T_b + S^e)\, [N^e]'\, dx, \; [A_4^e]_{p \times p} = \int_e \rho c\, [N^e]'\, [N^e]\, dx$$

$$[A_5^e]_{p \times p} = h\, [N^e]'\, [N^e], \; [A_6^e]_{p \times 1} = (LE - hT_a)\, [N^e]'$$

where

$$[B^e]_{l \times p} = \begin{cases} \left[\begin{array}{cc} \frac{\partial N_i}{\partial x} & \frac{\partial N_j}{\partial x} \end{array}\right]; \; p = 2 \\[2ex] \left[\begin{array}{ccc} \frac{\partial N_i}{\partial x} & \frac{\partial N_j}{\partial x} & \frac{\partial N_k}{\partial x} \end{array}\right]; \; p = 3 \end{cases}$$

$$p = \begin{cases} 2; \text{ for linear elements} \\ 3; \text{ for quadratic elements}; \end{cases}$$

$$[B^e] = \left[\begin{array}{cc} -\frac{1}{l(e)} & \frac{-1}{l(e)} \end{array}\right] \text{ for linear element}$$

$$[N] = \left[\begin{array}{cc} 1 - \frac{x}{l(e)} & \frac{x}{l(e)} \end{array}\right]$$

$$[B^e] = \left[\begin{array}{ccc} \frac{4x}{l(e)2} - \frac{3}{l(e)} & \frac{4}{l(e)} - \frac{8x}{l(e)2} & \frac{4x}{l(e)2} - \frac{1}{l(e)} \end{array}\right] \text{ for quadratic element}$$

15.6 Assembly of Elements

The region under study has been divided into n_e elements using n_n nodes. All the elements are assembled to get integral I as follow [23–26]

$$I = \sum_{e=1}^{n_e} I^e \qquad (15.11)$$

Where

n_e=22 for Model-1(A) and 1(B)

n_n=23 for Model-1(A) and 45 for Model-1(B)

Extremising I , we get

$$\left[\frac{dI}{\overline{\overline{T}}}\right]_{n_n\times 1} = \sum_{e=1}^{n_e} [D^e]^1_{n_n\times p}\left[\frac{dI^e}{d\overline{\overline{T}}^e}\right]_{p\times 1} = 0 \qquad (15.12)$$

Here

$$\frac{dI}{d\overline{\overline{T}}} = \left[\begin{array}{cccc} \frac{\partial I}{\partial T_1} & \frac{\partial I}{\partial T_2} & \cdot & \cdot & \cdot & \frac{\partial I}{\partial T_{n_n}} \end{array}\right]', \overline{\overline{T}} = [T_1\ldots\ldots T_{n_n}]'$$

$[D\ ^e]'$ shows the transpose of matrix $[D^e]$

$$[D^e] = \begin{cases} \begin{bmatrix} 0 & 0 & 0 & 1 & 0 & 0 & 0 \\ 0 & 0 & 0 & 0 & 1 & 0 & 0 \end{bmatrix}_{p\times n_n} & p = 2 \\[2em] \begin{bmatrix} 0 & 0 & 1 & 0 & 0 & 0 & 0 \\ 0 & 0 & 0 & 1 & 0 & 0 & 0 \\ 0 & 0 & 0 & 0 & 1 & 0 & 0 \end{bmatrix}_{p\times n_n} & p = 3 \end{cases}$$

15.7 Matrix Form of Element Equation

Matrix derivation of equation (15.10)

$$[A_1^e]_{p\times p} = \int_e K^e\,[B^e]'\,[B^e]\,dx = \frac{k}{l^e}\begin{bmatrix} 1 & -1 \\ -1 & 1 \end{bmatrix}$$

$$[A_2^e]_{p\times p} = \int_e M^e\,[N^e]'\,[N^e]\,dx = \frac{M^e l^e}{6}\begin{bmatrix} 2 & 1 \\ 1 & 2 \end{bmatrix}$$

$$[A_3^e]_{p\times1} = \int_e (M^e T_b + S^e) [N^e]' \, dx = (M^e T_b + S^e) \begin{bmatrix} 1 \\ 1 \end{bmatrix}$$

$$[A_4^e]_{p\times p} = \int_e \rho c \, [N^e]' \, [N^e] \, dx = \frac{\rho c l^e}{6} \begin{bmatrix} 2 & 1 \\ 1 & 2 \end{bmatrix}$$

$$[A_5^e]_{p\times p} = h \, [N^e]' \, [N^e] = \begin{bmatrix} 0 & 0 \\ 0 & 1 \end{bmatrix}, \text{``} Plank \text{ 's constant''}$$

$$[A_6^e]_{p\times1} = (LE - hT_a) \, [N^e]' = (LE - hT_a) \begin{bmatrix} 0 \\ 1 \end{bmatrix}$$

15.8 Algorithm and Computer Program to Solve Heat Equation using FEM in Matlab:

1. Set the value of constants/parameters
   - K, M, h,L,ρ,Tb,S, E, etc

2. Set Number of elements and time length
   nel=22;(Numbers of elemnts)
   nnel=2;
   ndof=1;
   nnode=23;
   sdof=nnode*ndof;
   deltt=0.1;
   stime=0.0;
   ftime=300;
   ntime=fix((ftime-stime)/deltt);

3. Set Mesh of the surface
   gcoord(1)=0.0;gcoord(2)=0.05;gcoord(3)=0.10;gcoord(4)=0.15;
   gcoord(5)=0.20;gcoord(6)=0.25;gcoord(7)=0.30;gcoord(8)=0.35;
   gcoord(9)=0.40;gcoord(10)=0.45;
   gcoord(11)=0.50;gcoord(12)=0.55;gcoord(13)=0.60;gcoord(14)=0.65;
   gcoord(15)=0.70;gcoord(16)=0.75;gcoord(17)=0.80;gcoord(18)=0.85;
   gcoord(19)=0.90;gcoord(20)=0.95;gcoord(21)=1.00;gcoord(22)=1.05;
   gcoord(23)=1.10;

4. Set Connectivity of the nodes
   nodes(1,1)=1;nodes(1,2)=2;nodes(2,1)=2;nodes(2,2)=3;nodes(3,1)=3;
   nodes(3,2)=4; nodes(4,1)=4; nodes(4,2)=5; nodes(5,1)=5; nodes(5,2)=6;
   nodes(6,1)=6;nodes(6,2)=7;

nodes(7,1)=7; nodes(7,2)=8; nodes(8,1)=8; nodes(8,2)=9;nodes(9,1)=9; nodes(9,2)=10;
nodes(10,1)=10; nodes(10,2)=11; nodes(11,1)=11; nodes(11,2)=12; nodes(12,1)=12; nodes(12,2)=13; odes(13,1)=13; nodes(13,2)=14; nodes(14,1)=14;nodes(14,2)=15;nodes(15,1)=15; nodes(15,2)=16; nodes(16,1)=16; nodes(16,2)=17;nodes(17,1)=17; nodes(17,2)=18; nodes(18,1)=18;nodes(18,2)=19;nodes(19,1)=19;nodes(19,2)=20; nodes(20,1)=20;nodes(20,2)=21;nodes(21,1)=21;nodes(21,2)=22; nodes(22,1)=22;nodes(22,2)=23;

5. Set Boundary and Initial Conditions

   bcdof(1)=1;bcval(1)=37;

6. Initialization of Vectors

   ff1=zeros(sdof,1) ;%initialization of system vector A3
   ff2=zeros(sdof,1) ; %initialization of system vector A6
   ff3=zeros(sdof,1);
   kk1=zeros(sdof,sdof) ; %initialization of system matrix
   kk2=zeros(sdof,sdof) ; %initialization of system matrix
   kk3=zeros(sdof,sdof);
   mm=zeros(sdof,sdof) ;%initialization of system matrix
   nn1=zeros(sdof,sdof);
   fn1=zeros(sdof,1) ;%initialization of effective system vector
   fsol1=zeros(sdof,1) ; %solution vector
   sol1=zeros(sdof,1) ; %vector containing time history solution
   kk1=zeros(sdof,sdof) ; %initialization of system matrix A1
   kk2=zeros(sdof,sdof) ; %initialization of system matrix A2
   kk3=zeros(sdof,sdof); %initialization of system matrix A4
   mm=zeros(sdof,sdof) ;%initialization of system matrix A5
   index=zeros(nnel*ndof,1); %initialization of index vector
   nn=zeros(sdof,sdof);
   index=zeros(nnel*ndof,1);

7. Distribution of Initial and Boundary Condition on different layers

   Layer -1 (Bottom Layer)
   for
   iel=1:10 (10 elements layer 1)
   nl=nodes(iel,1); nr=nodes(iel,2);
   xl=gcoord(nl); xr=gcoord(nr);
   le=xr-xl;
   index=feeldof(iel,nnel,ndof);
   k1=(k/le)*[1 -1; -1 1]; %A1

```
k2=(m*le)*[2/6 1/6; 1/6 2/6]; %A2
k3=(s*le)*[2/6 1/6; 1/6 2/6];%A4
f1=(Tb*m*le)*[1/2;1/2]; %A3
f2=(s1*le)*[1/2;1/2];*%A6
f3=[0;0];
right=(rho*cbar*le)*[2/6 1/6; 1/6 2/6];
[kk1]=feasmbl2(kk1,k1,index);
[kk2]=feasmbl2(kk2,k2,index);
[kk3]=feasmbl2(kk3,k3,index);
[ff1]=feasmbl3(ff1,f1,index);
[ff2]=feasmbl3(ff2,f2,index);
[ff3]=feasmbl3(ff3,f3,index);
mm=feasmbl1(mm,right,index);
    end
    Layer -2 (Middle Layer)
    for iel=11:18
nl=nodes(iel,1); nr=nodes(iel,2);
xl=gcoord(nl); xr=gcoord(nr);
index=feeldof(iel,nnel,ndof);
le=xr-xl;
a02=(k11*0.9-k31*0.5)/0.4;
a12=(k31-k11)/0.4;
b01=(m1*.9-m3*.5)/0.4;
b12=(m3-m1)/0.4;
c01=(s1*.9-s3*.5)/0.4;
c12=(s3-s1)/0.4;
A=[1 -1; -1 1];
C1=[ 1/3,1/6;1/6,1/3];
C2=[xl/4+xr/12 xl/12+xr/12;xl/12+xr/12 xl/12+xr/4] ;
D1=[1/2;1/2];
D2=[xl/3+xr/6;xl/6+xr/3];
k1=((a02/le)*A)+((a12/(le*2))*A*(xr+xl));
k2=((b01*le)*C1)+(b12*le*C2);
k3=((c01*0*le)*C1)+(c12*0*le*C2);
f1=((Tb*b01*le)*D1)+((Tb*b12*le)*D2);
f2=((c01*le)*D1)+((c12*le)*D2);
f3=[0;0];
right=(rho*cbar*le)*[2/6 1/6; 1/6 2/6];
    [kk1]=feasmbl2(kk1,k1,index);
```

```
    [kk2]=feasmbl2(kk2,k2,index);
        [kk3]=feasmbl2(kk3,k3,index);
    [ff1]=feasmbl3(ff1,f1,index);
    [ff2]=feasmbl3(ff2,f2,index);
    [ff3]=feasmbl3(ff3,f3,index);
    mm=feasmbl1(mm,right,index);
    end
```

Layer -3 (Top layer)

```
    for iel=19:21
nl=nodes(iel,1); nr=nodes(iel,2);
xl=gcoord(nl); xr=gcoord(nr);
le=xr-xl;
index=feeldof(iel,nnel,ndof);
    k1=(k31/le)*[1 -1; -1 1];
    k2=(0*le)*[2/6 1/6; 1/6 2/6];
    k3=(2*0*le)*[2/6 1/6; 1/6 2/6];
    f1=((Tb*0*le)* [1/2;1/2]);
    f2=(Tb*0*le)*[1/2;1/2];
    f3=[0;0];
    right=(rho*cbar*le)*[2/6 1/6; 1/6 2/6];
        [kk1]=feasmbl2(kk1,k1,index);
        [kk2]=feasmbl2(kk2,k2,index);
        [kk3]=feasmbl2(kk3,k3,index);
        [ff1]=feasmbl3(ff1,f1,index);
        [ff2]=feasmbl3(ff2,f2,index);
        [ff3]=feasmbl3(ff3,f3,index);
        mm=feasmbl1(mm,right,index);

    end
    for iel=22
nl=nodes(iel,1); nr=nodes(iel,2);
xl=gcoord(nl); xr=gcoord(nr);
index=feeldof(iel,nnel,ndof);
le=xr-xl;
    k1=(k31/le)*[1 -1; -1 1];
    k2=(0*le)*[2/6 1/6; 1/6 2/6];
    k3=h*[0 0;0 1];
    f1=((Tb*0*le)* [1/2;1/2]);
    f2=(Tb*0*le)*[1/2;1/2];
    f3=((L*E-h*Ta)*[0; 1]);
```

```
        right=(rho*cbar*le)*[2/6 1/6; 1/6 2/6];
            [kk1]=feasmbl2(kk1,k1,index);
            [kk2]=feasmbl2(kk2,k2,index);
            [kk3]=feasmbl2(kk3,k3,index);
            [ff1]=feasmbl3(ff1,f1,index);
            [ff2]=feasmbl3(ff2,f2,index);
            [ff3]=feasmbl3(ff3,f3,index);
            mm=feasmbl1(mm,right,index);
        end
        for in=1:sdof (Boundary Condition)
        fsol(in)=37;
            end
for i=1:nnode
sol(i,1)=fsol(i);
end
for it=1:ntime
tj=(stime+it*deltt);
tk=(stime+(it+1)*deltt);
fn1=deltt*(ff1+ff2-ff3)+(mm-(kk1+kk2+kk3)*deltt)*fsol1;
[nn,fn]=feaplyc2(nn,fn,bcdof,bcval);
fsol=nn\fn;
for i=1:nnode
sol(i,it+1)=fsol(i);
end
end
time=0:deltt:ntime*deltt;
propertyeditor('on')
plot(time,sol);
plot(time,sol(1,:),'–r',time,sol(10,:),'–r',time,sol(16,:),'–
xlabel('Time')
ylabel('Temperature')
```

Sub Programs

- **feeldof**

```
function [index]=feeldof(iel,nnel,ndof)
% purpose
% compute system dofs associated with each element
% %[index]=feeldof(nd,nnel,ndof)
% iel= element number whose system dofs are to be determined
```

```
% index=system dof vector associated with element iel
% nnel=number of nodes per element
% ndof- number of dofs per node
edof=nnel*ndof;
    start=(iel-1)*(nnel-1)*ndof;
    for j=1:ndof
        for i=1:edof
        index(i)=start+i;
        end
    end
```

- **feasmbl1**

```
function [mm]=feasmbl1(mm,right,index)
% purpose
% assembly of element matrix into the system matrix
% mm-system matrix
% m-element matrix
% index-dof vector associated with an element
edof=length(index);
for i=1:edof
    ii=index(i);
    for j=1:edof
        jj=index(j);
        mm(ii,jj)=mm(ii,jj)+right(i,j);
    end
end
```

- **feasmbl2**

```
function [kk1]=feasmbl2(kk1,k1,index)
% purpose
% assembly of elements matrices into the system matrix and assembly
of
% element vectors into the system vectors
% synopsys [kk,ff]=feasmbl2(kk,ff,k,f,index)
% kk-system matrix
% ff-system vetors
% k-element matrix
% f-element vectors
% index - dof vector associated with an element
edof=length(index);
```

```
for i=1:edof
    ii=index(i);
        for j=1:edof
        jj=index(j);
        kk1(ii,jj)=kk1(ii,jj)+k1(i,j);
    end
end
```

- **feasmbl3**

```
function [ff1]=feasmbl3(ff1,f1,index)
% purpose
% assembly of elements matrices into the system matrix and assembly of
% element vectors into the system vectors
% synopsys [kk,ff]=feasmbl2(kk,ff,k,f,index)
% kk-system matrix
% ff-system vetors
% k-element matrix
% f-element vectors
% index - dof vector associated with an element
edof=length(index);
    for i=1:edof
        ii=index(i);
        ff1(ii)=ff1(ii)+f1(i);
    end
```

15.9 Result and Discussion

Results obtained here are based on physiological facts. Variations in temperature distribution in the tissues are observed during the wound healing process. The Finite element method is more efficient as it covers the geometries of the region. The Heat distribution model can be further developed to study the interesting relationship among various parameters of the body region to understand the thermal changes caused in the process. The information obtained from this model can be validated by collecting real data with a machine learning approach [17–30] that will help biomedical scientists for application in the treatment of various diseases and help to develop protocols for medical purposes. This study is also helpful for the evaluation of the effectiveness of hyper-thermic treatments. The assessments of the danger

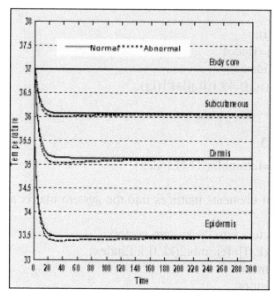

Figure 15.4   Graph between temperatures T and time t for $T_a = 15\,^\circ$C, E=0.0 gm/cm^2-min and $\eta=\theta$=0.01.

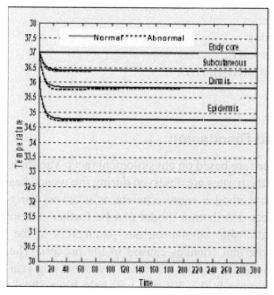

Figure 15.5   Graph between temperatures T and time t for T_a=15$\,^\circ$C, E=0.0 gm/cm^2-min and $\eta=\theta$=0.01.

involved and heat dissipation rate in soft tissues can be made. It may also help in investigations of thermoregulatory mechanisms. In the present model, various physiological parameters are taken for the calculation. The article is useful to solve any thermal distribution problem and the MATLAB codes are useful to develop and solve the mathematical model with small alterations in the codes. One can change the value of the parameters as per their need to develop and solve the model.

References

[1] E. A. Divo and A. J. Kassab, Boundary element methods for heat conduction: with applications in non-homogeneous media, WIT Press, Southampton (2003).

[2] J. Sladek, V. Sladek and S. N. Atluri, Local boundary integral equation (LBIE) method for solving problems of elasticity with nonhomogeneous material properties Computational mechanics, 24 (2000), 456-462

[3] S. Z. Feng, X. Y. Cui and A. M. Li, Fast and efficient analysis of transient nonlinear heat conduction problems using combined approximations method, International Journal of Heat Mass Transfer, 97 (2016), 638-644

[4] K. Yang, Geng-Hui Jiang, Hao-Yang Li, Zhi-bo Zhang, Xiao-Wei Gao, Element differential method for solving transient heat conduction problems, International Journal of Heat Mass Transfer, 127 (2018), 1189-1197.

[5] C. A Santos, J. M. V Quaresma, A Garcia, Determination of transient interfacial heat transfer coefficients in chill mold castings, Journal of Alloys and Compounds, 319 (2001), 174-186.

[6] Makrariya, Akshara, and N. Adlakha. "Thermographic pattern's in women's breast due to uniformly perfused tumors and menstrual cycle." Commun. Math. Biol. Neurosci. 2019 (2019): Article-ID.

[7] Makrariya, Akshara, and K. R. Pardasani. "Numerical simulation of thermal changes in tissues of woman's breast during menstrual cycle in different stages of its development." International Journal of Simulation and Process Modelling 14.4 (2019): 348-359.

[8] Makrariya, Akshara, and K. R. Pardasani. "Finite element model to study the thermal effect of cyst and malignant tumor in women's breast during menstrual cycle under cold environment." Advances and Applications in Mathematical Sciences 18.1 (2018): 29-43.

[9] Patil, H. M., and Maniyeri, R. (2019). Finite difference method based analysis of bio-heat transfer in human breast cyst. Thermal Science and Engineering Progress, 10, 42-47.

[10] Aziz, R., et al. "Artificial neural network classification of microarray data using new hybrid gene selection method." International Journal of Data Mining and Bioinformatics 17.1 (2017): 42-65.

[11] W. Q. Tao, Y. L. He, Q. W. Wang, Z. G. Qu and F. Q. Song, A unified analysis on enhancing single phase convective heat transfer with field synergy principle, International Journal of Heat Mass Transfer, 45 (2002), 4871-4879.

[12] Y. B. Tao, Y. L. He, Y. K. Liu, W. Q. Tao, Performance optimization of two-stage latent heat storage unit based on entransy theory, International Journal of Heat Mass Transfer, 77 (2014), 695-703.

[13] J. Ettrich, A. Choudhury, O. Tschukin, E. Schoof, A. August and B. Nestler, Modelling of transient heat conduction with diffuse interface methods, Modelling and Simulation in Materials Science and Engineering, 22 (2014).

[14] B. Deconinck, B. Pelloni, and N. Sheils, Non-steady-state heat conduction in composite walls, Proceedings: Mathematical, Physical and Engineering Sciences, 470 (2014) 1-22.

[15] J. Taler and P. Ocłoń, Finite Element Method in Steady-State and Transient Heat Conduction. In: Hetnarski R. B. (eds) Encyclopedia of Thermal Stresses. Springer, (2014).

[16] J. N. Reddy, Introduction To Finite Element Method | 4th Edition Paperback – 1 November 2020, Mc Graw Hills Publication , 4th Edition.

[17] W. McGuinness, et al., Influence of Dressing Changes on Wound Temperature. Journal of Wound Care, 13(9) (2004), 383-385.

[18] Makrariya, Akshara, N. Adlakha, and S. K. Shandilya. "3D Spherical— Thermal Model of Female Breast in Stages of Its Development and Different Environmental Conditions." Mathematical Modeling, Computational Intelligence Techniques and Renewable Energy: Proceedings of the First International Conference, MMCITRE 2020. Springer Singapore, 2021.

[19] Makrariya, Akshara, N. Adlakha, and S. K. Shandilya. "3D Spherical— Thermal Model of Female Breast in Stages of Its Development and Different Environmental Conditions." Mathematical Modeling, Computational Intelligence Techniques and Renewable Energy: Proceedings of the First International Conference, MMCITRE 2020. Springer Singapore, 2021.

[20] Makrariya, Akshara, and N. Adlakha. "Thermal stress due to tumor in periphery of human breast." J Biosci Bioeng 2 (2015): 50-59.

[21] Ali, B., Hussain, S., Nie, Y., Hussein, A. K. and Habib, D., 2021. Finite element investigation of Dufour and Soret impacts on MHD rotating flow of Oldroyd-B nanofluid over a stretching sheet with double diffusion Cattaneo Christov heat flux model. Powder Technology, 377, pp.439-452.

[22] Qiu, L., Wang, F., and Lin, J. (2019). A meshless singular boundary method for transient heat conduction problems in layered materials. Computers and Mathematics with Applications, 78(11), 3544-3562.

[23] Makrariya, Akshara, and N. Adlakha. "Quantitative study of thermal disturbances due to nonuniformly perfused tumors in peripheral regions of women's breast." Cancer informatics 16 (2017): 1176935117700894.

[24] Yu B, Cao G, Huo W, Zhou H, Atroshchenko E. Isogeometric dual reciprocity boundary element method for solving transient heat conduction problems with heat sources. Journal of Computational and Applied Mathematics. 2021 Mar 15;385:113197.

[25] Makrariya, Akshara, and K. R. Pardasani. "Numerical study of the effect of non-uniformly perfused tumor on heat transfer in women's breast during menstrual cycle under cold environment." Network Modeling Analysis in Health Informatics and Bioinformatics 8.1 (2019): 9.

[26] Sherrate, J. A. and Murray, J. D., ,"Mathematical analysis of a basic model for epidermal wound healing," Journal of Mathematical Biology, 29(1991), pp. 389-404

[27] W. Perl, An Extension of the Diffusion Equation to Include Clearance by Capillary Blood Flow, Annals of the New York Academy Sci., 108 (1963), 92.

[28] Aziz, R., Verma, C., and Srivastava, N. (2015). A weighted-SNR feature selection from independent component subspace for nb classification of microarray data. Int J Adv Biotec Res, 6, 245-255.

[29] Aziz, R., Verma, C., and Srivastava, N. (2015). A weighted-SNR feature selection from independent component subspace for nb classification of microarray data. Int J Adv Biotec Res, 6, 245-255.

[30] Musheer, R. Aziz, C. K. Verma, and N. Srivastava. "Novel machine learning approach for classification of high-dimensional microarray data." Soft Computing 23.24 (2019): 13409-13421.

Index

A

Aboodh transform 167, 168, 169, 176

Advection diffusion equation 211, 212, 214, 222

Alzheimer's disease 41, 49, 64, 210, 222

B

Basic reproduction number 181, 184, 196, 200, 205

Biological databases 64

Biological System 120, 271, 274

Boundary problems 271

Burger's equation 250, 252

C

Calcium buffers 41

Calcium 41, 44, 54, 56, 209, 212, 220

Chemical reaction 155, 156, 262

Cold environment 261, 267, 291

Computational convergence rate 241

Computational methods 63

Cone/wedge 156, 157

COVID-19 101, 102, 104, 113

D

De-Fuzzifier 101, 112

E

Endoplasmic reticulum 41, 44, 46, 53, 57

Equilibrium points 123, 179, 200, 201, 202

F

Finite difference method 49, 156, 211, 247, 249, 253

Finite element method 43, 211, 263, 271

Finite element technique 41, 44, 48, 57

Fractional calculus 4, 121, 209, 214, 221

Fuzzy transform 227, 230

G

Genetic algorithm 23, 25, 28

Genetic bee colony 23, 26, 28

Groundwater recharge 247, 249, 257

H

(HAM) with non-homogeneous term 155

Hartmann number 159, 164

High-resolution 227, 235, 239

Homotopy perturbation method 167, 169, 176

Hybrid differential 247, 252, 253

I

Independent component analysis 23, 37

Intracellular Fluid Crystallises 261

L

Laminar 137, 140, 158

Laplace transform 5, 209, 211, 212, 249

M

Mamdani fuzzy inference system 101, 103

MATLAB 41, 44, 101, 110, 200, 247

Micro-circulatory apparatus 262

Molecular docking 64

N

Naïve Bayes 24, 28, 78

Neurodegenerative disease 41, 209, 210, 216, 222

Newell–whitehead–segel equation (NWSE) 167

Nonlinear differential equation 167, 176, 239, 252

Non-linear PDE 137

O

Optimal control problem 121, 181, 196

P

Partial differential equation 44, 137, 245, 247, 271, 274

Particle swarm optimization 23, 38

PPI 63, 73, 93

Prandtl number 137, 147, 159

proteins 42, 63, 93

Pseudo-plastic fluid 137, 152

R

Risk Factors 101, 115, 116

S

Sensitivity analysis 181, 196

Singular differential equations 227

Spreading 102, 247, 249

Spreading Burger's equation 247

Stability analysis 8, 121, 123, 179, 181

Support vector machine 24, 29, 74, 94

T

Transient equation 271

Triangular base 227, 229, 230

Tumor growth model 227

V

Vertical jet 137, 138, 141, 143, 150

Voltage gated calcium channel 41, 42, 44, 45

W

Williamson fluid 155, 156, 157, 164, 165

About the Editors

Akshara Makrariya has a PhD in Mathematics from the National Institute of Technology (SVNIT) Surat, India, specializing in computational and bio mathematics. She completed her Post-Doctoral Research under the Young Scientist Fellowship (NPDF) programme, from the Science & Engineering Research Board (SERB), DST, and Government of India. Her research interest includes computational models to study temperature distribution in extended spherical and ellipsoidal shaped human organs, computational models to study thermal disturbances due to tumours, and their modelling with different numerical methods like fractional calculus, and the finite volume method, and by using optimization techniques.

Brajesh Kumar Jha obtained his PhD in Mathematics in the year 2013 from the National Institute of Technology (SVNIT) Surat, India. His research area includes mathematical modelling and simulation of calcium dynamics in astrocytes, finite element modelling of biological problems, and fractional differential equations and their applications in physiological problems.

Rabia Musheer has a Doctorate in Applied Mathematics from Maulana Azad National Institute of Technology, Bhopal, India. Her research interests include mathematical machine learning and data science application of mathematical modelling and optimization, soft computing, big data and bioinformatics.

Anant Kant Shukla a Doctorate from the VIT Vellore and CSIR-Fourth Paradigm Institute, Bengaluru. His research is in the area of analytical solutions and nonlinear systems.

Amrita Jha obtained his PhD in Mathematics in the year 2013 from National Institute of Technology (SVNIT) Surat, India. Her research area includes mathematical modeling and simulation of calcium dynamics in neuron cell, and finite element modelling of biological problems.

Parvaiz Ahmad Naik is working as Assistant Professor in Mathematics at the School of Mathematics and Statistics, Xi'an Jiaotong University P. R. China. He received his PhD in Applied Mathematics (Mathematical Biology) in 2015. His interest areas include differential equations, mathematical biology, fractional calculus, calcium dynamics, transmission dynamics of infectious diseases, etc.